Motherhood,
Rescheduled

THE NEW FRONTIER OF
EGG FREEZING
• *AND* •
*THE WOMEN WHO
TRIED IT*

SARAH ELIZABETH RICHARDS

Simon & Schuster

New York   London   Toronto   Sydney   New Delhi

Names and identifying characteristics have been changed.

Simon & Schuster
1230 Avenue of the Americas
New York, NY 10020

First Simon & Schuster hardcover edition May 2013

SIMON & SCHUSTER and colophon are registered
trademarks of Simon & Schuster, Inc.

For information about special discounts for bulk purchases,
please contact Simon & Schuster Special Sales at 1-866-506-1949
or business@simonandschuster.com.

The Simon & Schuster Speakers Bureau can bring authors to your
live event. For more information or to book an event, contact the
Simon & Schuster Speakers Bureau at 1-866-248-3049 or visit
our website at www.simonspeakers.com.

Designed by Nancy Singer

Manufactured in the United States of America

10 9 8 7 6 5 4 3 2 1

Library of Congress Cataloging-in-Publication Data

Richards, Sarah Elizabeth.
    Motherhood, rescheduled : the new frontier of egg freezing and the women
who tried it / Sarah Elizabeth Richards.—First Simon & Schuster hardcover
edition.
        pages cm
1. Pregnancy in middle age—Miscellanea. 2. Childbirth in middle
age—Miscellanea. 3. Reproductive technology—Miscellanea. I. Title.
    RG556.6.R53 2013
    618.20084'4—dc23   2013000700

ISBN 978-1-4165-6702-8
ISBN 978-1-4165-6729-5 (ebook)

For my mother

clock ticker, *noun.*
A female in her mid- to late thirties who must find a way to take advantage of her remaining fertility before she loses the chance to become a mother.

She is scared,
Scared to run out of time

                            Bonnie Raitt, "Nick of Time"

Contents

Motherhood,
Rescheduled

In the recovery room of a fertility clinic in midtown Manhattan, I popped open my eyes and wondered when my surgery would start. When I contracted my lower abdomen and felt my freshly tender ovaries, I realized it was already over.

"Am I done?" I called out to the nurses' station.

"Yes, you're done," a nurse responded. She handed me a Post-It note with the number 9 underlined twice.

"Is that the number of eggs?" I asked nervously. She nodded. As she left in search of Saltines and apple juice, I took a deep breath and felt a wave of relief wash over me. "They got them," I said to myself. They got them out of me while I was still thirty-six. No matter what happened from now on—infertility, romantic disappointments, ambivalence—I had some good eggs stashed away.

The nurse handed me post-op instructions: No driving for the day. No aerobics or sex for two weeks. Call if I have a fever, nausea, abdominal pain, or heavy bleeding. Ten minutes later, I floated down the hallway to the locker room, where I flung off my hospital gown, booties, and cap and pulled my polo shirt and shorts over my bloated belly. I hoped my boyfriend would be waiting with coffee; I was dying for my first morning sip. He smiled when I emerged into the waiting room.

I introduced Paul to the woman at the front desk so I could be released. I took the coffee in one hand and his arm in the other as we passed the couples in the waiting room. They were here because they needed medical help to make a baby. I was here because I was still fertile.

I let Paul navigate the lights as we walked across Park Avenue. Along the way, I saw mothers strolling with babies strapped to their chests, fathers walking sons to the park, nannies steering toddlers on bikes. For the first time in a long time, I felt nothing for them—not the usual pang that twisted my stomach, not the dread that pounded my head, not the ache that seized my shoulders.

I just felt hungry. I stepped up my pace and asked Paul what we should have for lunch.

Introduction

There was one time in my life when I was grateful for the biological clock. I was thirty-two years old and summoning the courage to leave a relationship. After nearly eight years of living with a man I deeply loved, I wasn't miserable. I just wasn't happy.

We had been doing the stuff the advice books say you're never supposed to do. We punished each other with silence, criticized each other's driving, made separate holiday plans, argued in public, and butted heads so fiercely about meals, budgets, sex, housework, exercise schedules, movie choices, and vacation destinations that it became easier to spend most of our free time apart.

We tried the standard fixes: we went to couples counseling, swapped lists of behaviors we were willing to change, and spoke in "I feel" statements. There were brief improvements, but the tension always returned, and I became increasingly certain that I did not want children with him. How could we agree on how to take care of another human being if we couldn't decide when to do laundry? I made sure never to miss my birth control pills.

I knew we had no future, but I also felt no urgency to overturn my life with a crushing, consuming breakup. There never seemed to be a good time, either. Who wanted to be alone during the holidays, the Fourth of July, the first day of fall? I would give it six more months, I told myself. Maybe we could read more books. Maybe we could try a new therapist. Maybe we could go on a long vacation. It wasn't *all* bad, I reminded myself.

Then that terrifying book came out. In the spring of 2002, Sylvia Ann Hewlett was detonating bombshells on nearly every talk show with *Creating a Life: Professional Women and the Quest for Children.* The message was clear: your fertility fades much sooner than you think; your eggs deteriorate dramatically after thirty-five and are pretty much fossils by your early forties. So listen up, all you clueless careerists! You've got to make having a family a priority. You'd better think twice about all your indulgent plans for advanced degrees, foreign postings, and after-work cocktails. Otherwise you're going to break your heart and the bank pursuing futile in vitro fertilization (IVF) treatments in an attempt to "snatch a child from the jaws of menopause." That's not to mention the increased risk of having a baby with Down syndrome *if* you manage to get pregnant.

I joined my generation in a collective gasp. "*Now?*" I whined to myself. I had just finished graduate school and was trying to launch my career as a freelance journalist. Plus, I still had to break up, grieve, find a new apartment, move out, lose ten pounds, acquire new relationship skills, and try to meet someone else. *Then* I had to get engaged, marry, and make a baby. That left very little contingency for rebounds, bad judgment, and trouble becoming pregnant.

If everything went as planned, I could have my first baby at thirty-seven and maybe fit in a second by thirty-nine. "My God!" I exclaimed to my girlfriend over the phone. "I've already lost my third child!"

Before Hewlett's book, I had assumed that I would be a mother, just as I knew I would marry, buy a home, and at some point fit into those Oshkosh B'Gosh short overalls I bought two sizes too small in college. I sleepily went about my life and took comfort in the pleasant stupor that was *someday*. I had little sense there was an actual deadline and that it was looming. Life was challenging enough without God suddenly setting a timer.

Without knowing it, I had become a Clock Ticker, and my pleasant stupor was replaced by the loud hum of the clichéd biological clock, which began to torment me like a clunky old air conditioner. My friends started having babies, and I was suddenly behind. I overheard my parcnts making excuses for me to their friends: *She's busy with her job. She's a late bloomer. She's picky.* In the most discouraging sign, relatives stopped asking when I planned to get married and start a family, as if I had been relegated to being the Crazy Aunt at family gatherings.

There were statistics to prove you were not alone, and that you were a member of a swelling demographic of women who had delayed marriage and motherhood. Supposedly one in five women was waiting to start her family until after age thirty-five, a percentage that had increased nearly eight-fold since 1970. And for the first time in history, more children were born to women over thirty-five than to teenagers.

You saw enough older new mothers in your neighborhood to know the statistic was true, but secretly you still wondered if there was something wrong with you because life hadn't worked out for you the way it had for Everyone Else. You told yourself that there was nothing wrong with being the Last One Left. You were simply on a different path and would make good decisions for your future. But you still felt a little twinge of sadness every time you saw your single-line listing on a family reunion attendance list. Or shopped for boxed Christmas cards of New York City snow scenes because it seemed ridiculous to write a holiday letter about yourself. Or realized that you were one of a few friends from high school holiday get-togethers available to go out after 8 p.m.

I wish I could say that I trusted everything would work out and that I carried myself with a *Secret*-like confidence that made me wildly attractive. I didn't. I spent the majority of my thirties alternately freaking out and talking myself down. I paid thousands of dollars for therapy, drank too much wine, and harassed my busy friends and family with distraught phone calls. I often repeated their encouraging words in my head before I went to sleep: *I still have time. There are lots of good guys out there. I am in a better place now to choose a mate than I was in my twenties. I have learned a lot of relationship lessons. I still have time.* But only one thing gave me real comfort. If I actually did run out of time, I had a list of motherhood options taped to my desk lamp: donor eggs, foreign and domestic adoption, other couples' leftover IVF embryos, stepchildren. I knew the alternatives came with their own complications, but I thought they were ones I could live with. And if I had to live with them in the same house as my fabulous new husband, all the better.

So why did I still feel so awful? When the panic subsided, I was gripped with sorrow that I was losing my chance to have a biological child. Not an abstract baby but the baby I had dreamed about my entire life. Her name had changed over the years from Ashley to Chloe to Claire, and she wore various outfits and hairstyles. But in my imagination, she was always three years old, clomping around in my high heels, freezing apple juice in ice cube trays, and saying things that made me go gooey inside, such as "This is the best day ever, Mommy."

I wondered why it was so important that she shared my DNA. Was I curious to see if she had my high forehead or loved mustard and rainy mornings like I did? Was I intrigued by the idea of creating an extension of myself and the man I loved? Or did I crave seeing several generations of my family in one photo? All I knew was that the idea of never meeting Claire (or her younger brother, Henry) seemed utterly tragic.

I can't remember when I first heard about egg freezing, a procedure that promised to make the biological clock obsolete. The concept was extraordinary. Hormone shots made your body pump out eggs, which were surgically extracted and frozen. When you were ready to become a mother, the eggs could be thawed and fertilized in a lab with your future partner's sperm to make embryos. Just as in standard IVF, those embryos would be placed inside your uterus to grow into babies. The only difference is that you would be using your younger, hardier eggs in case your current eggs were no longer viable. And since it's possible for women to carry a baby well into middle age, you technically could become a mother whenever you wanted. Not that you would necessarily want to have a baby at, say, fifty or sixty, but the point was that you *could*.

When egg freezing first appeared in the cultural cosmos early in the past decade, my friends and I talked about it in a hyperbolic way, as when a seventh-grader frustrated with boys shouts, "Fine! I'll just become a nun!" We loved this handy metaphor to communicate our fears about becoming lonely spinsters. "Well, at this rate, I'll have to freeze my eggs," we might say in the same tone as "move to Alaska" (a state with a heavily skewed male-female ratio). We threw it around as cheap drama intended to be oddly reassuring because, after all, who would really have to resort to something so expensive and extreme?

It was only after I turned thirty-five—the point of no return—that I began to seriously consider what egg freezing could do for me. I fantasized about what it would be like to be free from the suffocating press of time that constantly reminded me that my entire future happiness depended on the decisions I made over the next few years. It meant I could escape the penalties for lacking the spine to leave a relationship I knew was not working. It meant redemption.

Egg freezing seemed too good to be true, in the same baffling way a diet pill promises to magically wipe out the caloric damage of waffle fries or Botox can make you look forever twenty-five. It felt unnatural and sort of unfair, as if I could buy the privilege to take the final exam later than the rest of the class or skip filing taxes. I wondered if I would approach dating differently if I was no longer in such a rush. Maybe I'd treasure these next few "free" years unencumbered by baby anxiety and write a bunch of books, train for a marathon, and solidify a marriage. That way, I'd be in an even better position to be a mom.

Would egg freezing clarify my life? Complicate it? Or leave me right where I started?

My girlfriends and I loved to linger over dinner and imagine what life would be like off the clock. We speculated that some of our married friends may have been more selective and would now enjoy better marriages if they hadn't feared being "put out to pasture." We lamented peers who had panicked prematurely and gotten stuck with reluctant fathers or left home alone as single moms. We wondered whether gun-shy friends would make more of an effort to date if they thought their ovarian age no longer mattered. We mused about what it would be like to have the same reproductive freedom as men, even to date and think like men. Or (gasp!) actually date *younger* men—a biologically attractive pairing, considering women live on average six years longer than men. Men would also have more choice of partners, since those who wanted children wouldn't have to seek out for their egg quality women who would never understand why they spent nights and weekends playing Dungeons & Dragons in high school.

In any case, surely men would appreciate a little less pressure from us. They also might be more willing to give relationships a serious try if they didn't fear they were wasting our last chance at motherhood.

However, the idea of stopping the clock also made us a little uneasy. You could never actually count on egg freezing, and any relief was over-shadowed by the awareness that your remaining fertile years were still ticking away. So far, the success rates of egg freezing ranged wildly and were always accompanied by asterisks explaining that the technology was rapidly improving and that you could game the odds by freezing several rounds of eggs. Still, it was a bewildering equation. How were you supposed to be cautious and hopeful at the same time?

Even if egg freezing did work, the question remained: Would tin-kering with such a finely tuned reproducing machine lead to harmful social and personal consequences? You would be an older mother and might endure a more difficult pregnancy. You might not see your chil-dren marry or know your grandchildren. Your own parents might not even meet their grandchildren. As my wise friend Janelle said, "Maybe you'd just drag out the whole thing. At forty-five, you'd still worry if you were ready to be a mom. Without the clock, there's no trigger to force you into action." She had a point. Deadlines serve a purpose in life.

But my friends and I could only speculate because, like most peo-ple who talked about egg freezing in 2005, we had not frozen our eggs or met anyone who had. As I considered the procedure, I wanted to meet other women who could tell me how egg freezing had affected their lives. Did it take off the pressure and help them relax more? Did they date differently? Marry later? Most important, did their frozen eggs help them have babies years later, when their natural fertility was gone? The journalist in me wanted to explore this medical breakthrough and changing social ideas about coupling, mating, and parenting. The Clock Ticker in me wanted to find a group of big sisters who could share how they navigated this difficult and confusing time.

However, in 2005 only a couple dozen women had frozen their eggs, let alone tried to thaw them. So I approached the handful of doctors who were offering the procedure at that time and asked to be put in touch with their earliest patients. Kevin Winslow, a fertility doctor in Jacksonville, Florida, and Megan Aitken, a patient educator at the national egg-freezing network Extend Fertility, introduced me to women who were willing to include their stories in a book. They are uncommonly generous women who spent countless hours with me sharing their innermost fears, anguish, and joys. I wanted them to feel comfortable speaking openly about their lives and past relationships, so I have changed their names and disguised their identities, as well as those of their acquaintances and loved ones.

I weave in the history of the science and controversy about whether egg freezing should be offered to healthy women by including two important figures on opposite ends of the debate: Dr. Michael Tucker, an embryologist who pioneered the procedure in the mid-1990s, and Christy Jones, an entrepreneur who promoted it to the masses. I also tell my own story.

And in the end, I attempt to answer the looming questions: What happens to women *after* they freeze? Did the act of freezing put them on a different life path? If so, was it a better one? Was egg freezing the reprieve we imagined?

I also hope to challenge the cultural stereotype of how Clock Tickers are viewed. There are many reasons why a woman might freeze her eggs. A 2010 meta-analysis of twelve studies identified the following ones: divorce, family history of premature menopause, career or educational opportunity, lack of supportive partner, or just not feeling financially secure or emotionally stable enough to bring children into their lives.

But the media clings to two caricatures: the materialistic career girl who is too ambitious to pay attention to her own biology (translation: she's too selfish and independent to be married) and the plucky single gal "still looking for Mr. Right" (translation: she's undatable or too picky).

Older women are often described as becoming mothers after battles with infertility. Stories frequently report the number of miscarriages they have suffered and that their "miracle" baby is a "long-awaited" reward for faith and persistence. But egg freezers are a different lot. They weren't caught off-guard by infertility; in many cases, they're *planning* on becoming older moms. They're even seen as delusional for putting their faith in such young and unproven technology. Or they are called entitled and immature (just like the woman in the "Oops! I forgot to have a baby" postcard). They frittered away their fertility drinking mango-tinis and climbing the corporate ladder while the rest of the world hunkered down and raised children, whether or not *they* were ready.

Even as society grows more comfortable with the new reproductive power of forever fertile fortysomethings, we're still not completely comfortable with the concept of egg freezing—or sympathetic to the women who freeze. If their half-baked baby plans don't work out, we quietly think they deserve their fate. Or if they do work out, we wonder if their kids will pay the ultimate price because their mothers messed with nature.

Why doesn't a woman who freezes her eggs ever get credit for proactively seeking a solution to her so-called problem of waiting to have a baby? Why isn't she celebrated for enduring hormone shots and emptying her bank account in order to have a better chance of finding a partner and father for her children, avoiding birth defects, and becoming financially secure so she can hold up her end of a marriage or support a family?

This "take charge" attitude is one of our most fundamental American values. Seen this way, egg freezing isn't an act of desperation or indulgence. Rather, it's an act of love for her future family.

1. *Longings*

Sarah

When I called my mother in San Diego with the news that I was ready to leave the man I had lived with and loved for the previous eight years, she flew to New York and collected me and my belongings. We rented a car and for the next four days drove back and forth from Manhattan to big box stores in New Jersey to buy furniture, dishes, and rugs for my new studio apartment. We painted the walls a cheery polenta color intended to soothe me during the dark nights ahead and show potential suitors that I could make a home and was Wife Material. It didn't hurt to think ahead, my mom insisted.

"Just tell me you're not going back!" she pleaded as she sank into an Ikea showroom couch. "Tell me this is all worth it."

She was right to be worried, considering I had spent the past couple of years trying to leave but never staying away for long. The failed breakups included months apart at job internships and in summer sublets. But this was different. I had my own apartment. I had a year-long lease. I also felt I didn't have a choice. I was nearly thirty-three and knew I had to become serious about finding a partner if I wanted marriage and kids. I couldn't waste any more time hoping my floundering relationship would improve.

But my anxiety about surviving a breakup was nothing compared to the terror I felt about entering the dating market, especially in notoriously competitive Manhattan. Except for a make-out session with my apartment broker, I hadn't dated in nearly a decade. The very thought of having to shop for date-night sweaters and toil through awkward conversations made me want to run back to the safety of the Barnes &

Noble Personal Growth section, where I had spent many a transformational evening on the floor nestled between Hardcover Mystery and Weddings with a stack of books.

I had never been very good at the dating game. Prior to meeting my ex, my relationship résumé included some college hookups interspersed with a couple of months-long flings with guys who liked girls like me, with preppy bobs, boxy black shoes, and smart mouths.

I had since grown out my hair and developed better taste in shoes. Still, I spent the next few months getting ready for battle. I lost weight and shopped for silk tops and butt-enhancing jeans. I bought highlights, self-tanners, and teeth whiteners. I experimented with lip gloss and eye shadow. I started going to a straight-talking therapist. I read a lot of dating advice books, which promised that by making myself busy and interesting I would naturally become a magnet for men fascinated by my "full life." So I joined college alumni clubs, professional organizations, hiking and running clubs, a softball team, and the New York Junior League. I attended wine tastings, university club mixers, lectures, and film festivals. I took spinning and yoga classes. I went on sailing, skiing, and diving trips.

My friends and family piled on encouragement. My mom suggested that my relationship had served as a long-term "cocoon" and that I would have an advantage over any jaded peers who had spent those years bouncing between boyfriends and in and out of bars. One friend said I could enjoy the pool of men coming off their first marriage who had made their mistakes and were ready to *really* settle down. A male friend told me over drinks that he loved dating women in their thirties. Every twenty-eight-year-old is all about getting married, he explained. "But chicks in their thirties, they've stopped caring as much. They're much more fun."

I stared at him, horrified. "Sounds great!" I replied. "I can't wait to get out there."

I really *couldn't* wait. Six months after my breakup, I was thirty-three years and three months old. I needed to make a plan. I did not want to let my entire reproductive future depend on chance encounters at Duane Reade or setups that my friends never got around to setting up. All of my guy friends, the ones who had gotten away, were now gone for good: They were married. I had to take the accelerated course; I had to go on-line. "It's fantastic," my girlfriend, a self-proclaimed "dinner whore" who had met her fiancé on Match.com, promised. "I got taken out so much I cut my grocery bill nearly in half."

I felt slightly nauseous when I submitted my first online dating profile to Match.com for approval on a Friday evening in the fall of 2003. But I was quickly heartened when I scanned the pages of prospective beaus. The site was full of cute, successful men who enjoyed fine dining, hiking the Inca Trail, making waffles on weekends, and posing for photos with their nieces and nephews.

I had made my best marketing effort. I followed the books' advice to write in my profile that I was easy to please and independent (important) but still needed a man in my life (more important). On a trip back home to San Diego, my mother filled a digital memory chip of photos of me wearing a red sweater in front of her pink bougainvillea. (Men like color, the books said.) She even looked up flattering posing positions on the Internet and instructed me to cock my head to the right so that I would seem approachable and fun. We tried another ensemble, a cream sweater and pearls, but ditched it when my cousin's boyfriend said I looked too much like a mom—ironically not the look I was going for. Next I tried on a black top and jean skirt and reclined on the lawn holding an empty Diet Coke so it appeared as if I was at a barbecue. (Men also like full-body pictures.) We then spent the afternoon printing and ranking the photos and asking friends and family members to weigh in, as if we were asking their opinions of tile samples for the kitchen.

"What are you, the Family Marriage Project?" my brother teased me. "I would never let Mom know I was on Match."

"She's just helping me with the pictures," I replied, feeling slightly embarrassed. But part of me didn't mind being the Family Marriage Project. It was better than the Still Unmarried Sister.

By Saturday morning my profile had been approved and was visible to other members. I held my breath as I logged into my account. When I discovered twenty new emails, I burst into tears of relief. Some I immediately deleted: the Parisian pharmaceutical rep who wanted a date to "show him the town" during his next business trip, an engineer looking for a "drug-free, disease-free, and drama free" woman, and another who included a picture of himself wearing nothing but a towel wrapped around his waist. But the windfall also included a lot of earnest guys who made thoughtful comments about my profile and knew where to find the best hot chocolate in the city.

The email volume soon leveled off, but I consistently received inquiries from men over forty-five. "They're married and looking for affairs," one friend concluded. "They want you to help them feel young again," offered another. But more often than not, this was their story: they had spent many years focusing on their career and now wanted a family. They claimed they were still "youthful," had a lot of energy, and could offer the right woman plenty of financial security. One fifty-two-year-old man who was visiting New York from my hometown of Encinitas, twenty-five miles north of San Diego, told me his thoughts outright over brunch. "You would be perfect," he said in between forkfuls of frittata. "I hope I haven't forgotten to mention that I would take very good care of you."

These men didn't care that I liked lazy Sunday mornings with the *New York Times*, beach bonfires, or spicy salmon rolls. They cared about my fertility. Even men my own age seemed anxious to start the process. "I need to know where this is going," one said after a couple months of lukewarm dates. "I'm at that point in my life where I want something permanent. I also want kids." I told him I needed time to see how the relationship evolved. He told me he needed to move on. What had happened to men's commitment problems I had heard so much about?

Despite my exuberance over finding myself in a buyer's market, I still felt uneasy that I was years away from catching up to my peers. My younger brother had just announced that his wife was pregnant. Even some of my single friends in their mid-thirties had become less than careful with birth control. One stopped reaching for the condoms mid-passion with a man she had been dating for five months. She became pregnant, and they were married shortly after.

Another allowed her boyfriend to "pull out" regularly during sex. She didn't use a backup form of birth control and figured pregnancy wasn't the worst way to compel him to commit to her. She ended up a single mother paying hefty day care bills. "Men come and go, Sarah," she reasoned. "But I'm bringing life into this world, and that's what really matters." Was I so late joining the game that I needed to go straight to babies and worry about finding a good marriage later? Or should I be looking for the "good enough" guy now?

Over the next couple of years, I went on a seemingly endless number of dates. I had lots of mini-relationships. In some, I became wildly infatuated and got dumped. In others, I plodded along with nice guys I could barely stand to kiss. Maybe one of those Match men would have turned out to be a wonderful husband, and I would now be settled in the suburbs enjoying the comfort of mom clogs and Saturday mornings with Claire at Costco. But I never fell in love with any of them. I knew without a doubt that I wanted to be a mother someday, but I didn't want it badly enough to overlook my heart.

As part of my quest to become a more highly effective person in the dating scene—and life—I came across Stephen Covey's mantra of "working with the end in mind" for setting goals. That way, he said, you didn't just let life happen to you: you steered it. But I didn't like the idea of making children the ultimate goal that inspired my search for love. Falling in love was one of life's great mysteries. I couldn't deal with the idea that it would be reduced to a bullet point on a reproductive plan.

In the fall of 2005, more than two years after I had become a serial online dater, I received an email through an online dating site called SquareDating.com from a forty-one-year-old man who called himself "NYCPaul." I scrolled through his profile and stopped at the question "Kids?" He had checked "Maybe."

I usually passed on anything other than "Definitely" or "Yes." But I was struck by NYCPaul's wide smile and kind eyes. "Maybe" wasn't "No," I thought. I wrote him back.

Monica

Monica Reyes was preparing a sales presentation for a client she was visiting in Minneapolis when she felt a sharp pain shoot through her chest. It spread to her neck and started to burn. Alarmed, the thirty-six-year-old product manager immediately called her father, who was an emergency room physician back home in New Jersey. "Dad, it feels like I'm having a heart attack," she whispered. He asked if she was dizzy. She wasn't. Short of breath? No, not that either. "You're not having a heart attack," he concluded and suggested she take some Motrin.

Monica didn't like traveling alone in a different time zone with her chest on fire. What had begun as a minor ache behind her rib cage was now agonizing. "Should I go to the hospital?" she pressed her dad.

"You don't need to," he insisted. She was relieved when she felt better a few hours later. Still, she made an appointment with her family doctor when she returned home to New Jersey. He surveyed her medical history, reviewed her dietary habits, and ordered a chest x-ray, electrocardiogram, MRI, and CAT scan. His diagnosis: acid reflux. His treatment: antacids.

Monica felt reassured until the pains surfaced again over the next couple of months. Sometimes they were strong and sudden; at other times they were dull and persistent. Lacking any other explanation, she chalked it up to stress. After all, her job consumed her, and she spent nearly half of her workdays on the road. She might be gone a week at a time; some days she flew into a city for a lunch meeting and returned home the same day. She was thoroughly exhausted. When she looked in the mirror in the harsh light of airport bathrooms, she wondered if her glazed eyes and tight jaw communicated hardness. If they did, the persona was a lie; underneath she felt fragile and lonely.

Monica didn't buy into all the self-help advice she had read about treasuring her single years. She hated being unattached and had been looking in vain for a boyfriend since breaking off her engagement a few years earlier. She had tried meeting men through the online dating site eHarmony as well as attending several speed-dating events, during which she was paired with nearly a dozen men in one evening, but she never felt a spark with any of them.

She wanted a companion and missed the feeling of being in love, but lately her search had taken on a new urgency. Out of the blue a year earlier, her desire for a baby had surfaced. The pangs hit hard and hurt physically, and Monica was as surprised as anyone that she could feel such profound longing. She wasn't one of those women who had always known she would be a mom. Although she had grown up in a traditional Filipino family, she had been a tomboy and hung out with her brother, who was ten months younger. She played field hockey and boys' Little League. She preferred Snoopy over Barbie, and her toys consisted mostly of arts and crafts supplies. During the major romances of her twenties and early thirties, she occasionally thought about having children, but her fantasies seemed plucked from magazines: babies as stylish accessories or hobbies for couples to share.

She first flirted with the idea of becoming a mom at age twenty-five, after she fell in love with an American man while working in Japan. She thought Albert would make an excellent father, since he had worked as a camp counselor and lifeguard when he was younger and had such an easy rapport with kids. They both liked to hike, and she imagined walking behind him cooing at their baby nestled in a baby backpack. They talked about how they'd like one boy and one girl. But three months into their relationship, Albert moved back to Texas. They tried to maintain a long-distance romance, and a year later he asked her to move back to the United States with the promise of an engagement ring. But when Monica looked for a position in Texas, she realized the time apart had taken its toll. She didn't feel connected to him. She received better job offers in New Jersey, and they broke up shortly afterward.

A few years later, in business school, she met a Danish classmate named Lars and followed him to Copenhagen after graduation. They also planned to marry and start a family. Monica fantasized about Lars caring for their infant during his generous Danish paternity leave. During the afternoons, they could stroll through the nearby park pushing a fancy English pram.

Their daily life was less romantic. Lars's work required him to travel constantly, and Monica, then thirty-one, sensed they were drifting apart. She focused on studying Danish, looking for a job, and making friends with other expat women, but within weeks she felt isolated and depressed. She also resented having to do most of the cooking and cleaning because Lars was always so busy. "I have an MBA," she fumed. "How did I get stuck doing all this?" She loved Lars dearly and wanted to make the relationship work, but she didn't know how to ease the tension, let alone talk about weddings. "If this is something you don't want," she said at yet another awkward dinner one evening, "you have to tell me."

Two weeks later, when Lars came home early from work one day with a doleful look in his eyes, she knew what was coming. "Monica, I can't do this anymore," he confessed. The next week, six months after arriving, she and her heavy heart boarded a plane back to Newark.

In the past, Monica's baby yearnings had been as fleeting as her wish for a new overseas work assignment. But at thirty-five, the desires became visceral and unrelenting. When she saw an attractive man wearing a Baby Björn carrier with an infant in it, her legs turned to jelly. At church, the beautiful, well-dressed families huddled together in pews made her eyes well up with tears, and she felt an unsettling surge of jealousy. During the service, she prayed hard. "O God, give me the strength to be happy for others who have what I don't have yet. Give me the strength to be patient and focus on what I need to be doing." But nearly every week she was overcome with emotion, until she stopped going altogether.

Monica hated feeling envious. If she wanted something—an MBA, a townhouse, a job abroad—she simply figured out a way to get it. Now she was tortured by a desire she couldn't satisfy. Her sister, who was eighteen months older, already had two kids in elementary school, and her younger brother was newly married and wanted to become a father soon. Her mother didn't ask her about her dating prospects, but Monica could tell the subject was on her mind. Her father, on the other hand, constantly worried about her being alone. He asked her to email him her travel itineraries and reminded her to turn off the water to her washing machine so the pipes wouldn't burst during freezing weather. Monica also was anxious about living by herself. What if she choked on a vitamin? Slipped in the tub and hit her head? Fell off a chair while reaching for a high shelf in her kitchen?

Even before the baby urge hit, Monica had been aware of her creeping age. When she became engaged at age thirty-three to Ben, a man six years older who wanted to start a family right away, she thought she had plenty of time to have several children. She had been happy that after all those romantic false starts, she was finally going to settle down. Unlike her other romances, though, she didn't dream about having a baby with Ben.

She wasn't having fantasies about planning her wedding, either. Usually when she made decisions, she became obsessive about following through with them. She told herself she was just overwhelmed with details. But when she accompanied Ben to sign the contract for the reception hall, she had to fake enthusiasm. She told herself she would feel like a real bride when she found her dress. But when she admired her reflection in a stunning silk gown, she couldn't visualize walking down the aisle.

"What is wrong with you?" she chided herself. "You should be happy about getting married." Maybe they had rushed into it, she thought. Ben had proposed five months after they had met at work. She didn't feel as infatuated with him as she had with her other boyfriends, but she told herself they had a more mature, comfortable kind of love. Maybe she was too independent and was put off by the way he had taken over much of the wedding planning. Maybe they just had clashing personalities. She was troubled by how frequently their bickering turned nasty. Once, a disagreement over how to negotiate with her boss escalated so quickly that Ben threatened to smash her Murano wine glass collection; she retaliated by warning she would throw his encased Philadelphia Phillies' Hall of Famer Mike Schmidt–autographed baseball across the room. She had never become so worked up before. Now she questioned how their marriage could last if their engagement was already so volatile.

"Maybe we need more time," she urged him in a bid to postpone the wedding. Ben immediately put her on the defensive.

"Why are you getting cold feet?" he pressed. "Even your family knows you're afraid to commit."

Was she? Did they? In an effort to figure out her feelings, Monica made a list of what was good about Ben: He would be an excellent provider. Her family liked him. They were good companions and both enjoyed playing golf, attending the opera, and traveling to Europe and Florida. He had even accompanied her family on a trip to the Philippines. Then she made a list of the bad: She didn't feel connected to him. She didn't like his abrasive personality. She hated their power struggles.

Then she wrote down the reservation she had been scared to admit to herself: she didn't love him.

On a chilly spring night a few weeks later, Monica could no longer ignore the growing collection of red flags. As she and Ben were licking the envelopes of the wedding invitations that were due to go out the next day, she felt an unmistakable sense of doom. Although it was near midnight, she grabbed her car keys and headed for the door. "I've just got to get out of here for a while," she announced and ran to her car.

As she drove aimlessly down the dark side street near the office park where they worked, she started to tear up. Soon she was bawling. She pulled over to the side of the road and called her mother. "I can't do it! I just can't," she cried.

"It's okay, honey," her mom consoled. "You don't have to."

Monica backed out of the wedding but stayed with Ben for two more months before ending the relationship for good. However, she didn't cancel the appointment she had made earlier with her priest for premarital counseling. She wanted reassurance she had made the right decision.

"Where's your fiancé?" the priest asked when she showed up alone.

Fighting back tears, she explained that she had called off the wedding because it didn't feel right. She was relieved when he praised her.

"I admire you because you're very brave," he said, explaining that many people ignore warning signs because they feel stuck on the wedding-planning train and don't want to disappoint their families. "But you jumped off."

Kelly

At age thirty-seven, Kelly Dunn was the most worrisome kind of Clock Ticker: the kind who didn't know she was one, the kind who had no plans to marry her boyfriend of two years.

Kelly never talked about *if* she would have a family; she just knew she would have one someday. She kept a giant file with decorating ideas for nurseries and ads for cute maternity and children's clothes that she tore out of magazines. On weekends she scoured antiques fairs for vintage baby dresses or little "John John" suits with Peter Pan collars. She even bought a christening gown from the mid-nineteenth century that she planned to pass down to her kids as a family tradition. At home, she stuffed three plastic comforter bags with relics from her own babyhood: quilts sewn by her mother, her silver hairbrush and spoon, and her favorite Golden books, *Little Red Riding Hood* and *The Night before Christmas*.

When the women in her office talked about their families, Kelly refused to be left out of the conversation. "I can't wait to have a big family," she'd say, often making outrageous claims about her future children. "My children will take piano lessons, love antiques, get good grades, volunteer in the community, and eat all their fruits and vegetables." If she heard about one of her coworker's kids misbehaving, she teased, "My kid would never throw a tantrum like that!" Her coworkers usually responded by rolling their eyes or chuckling. Kelly used the strategy to deflect their sympathy over the obvious fact that she didn't have a family.

When her friends' kids visited the hotel near Charlotte, North Carolina, where Kelly worked as the director of sales, they all made a beeline to her office—and the special drawer at the bottom of her filing cabinet, filled with promotional items she collected at trade shows: flashlights, clickers, puzzles, stress balls. They were each allowed to pick one prize to take home. Sometimes they sat in her lap and watched a funny video she showed them on the computer. Their parents would often watch from her office door. "You'd be such a good mom," they would say.

Still, she was surprised when her coworker, Kathy, cornered her one afternoon in the fall of 2002. "I'm not trying to manage your personal business," Kathy began, "but I know you want a baby. I wanted to tell you what I just heard about this thing called egg freezing."

Kathy explained what she had learned during her latest trip to a fertility clinic in her hometown of Jacksonville, Florida, where she was undergoing IVF treatment while visiting her parents. "I first thought about it for my younger sister, but on the flight back, I thought it would be perfect for you too," she said. "It could give you more time."

Kelly politely thanked Kathy for the information but wondered why she talked about it so urgently. Kelly was just thirty-seven; she had plenty of years left to have kids. She knew getting pregnant could be harder as she grew older. Kathy, for instance, was undergoing IVF at age thirty-eight. But she had read lots of stories about women in their forties becoming moms: Julianne Moore at forty-one, Madonna at forty-three; Geena Davis at forty-six. "If they can, I can," she thought. Besides, she was healthier and more energetic than most people her age. She ran six miles a day and lifted weights several times a week. She didn't smoke and rarely drank. She ate lots of fruits and vegetables. She was blessed with her mother's beautiful unlined skin and flat stomach. Surely her ovaries were in good shape too.

Her family didn't seem too worried about her age, either. Then again, Kelly was the youngest of four children and had twelve nieces and nephews, so there never seemed to be a lack of kids around. Or maybe they were just being kind because they knew how much she wanted to be a mother and how heartbroken she had been when her marriage had ended a few years earlier.

Kelly had met the man who would become her husband a few years after college, and it seemed as though they were always moving around. He worked long hours and traveled frequently, trying to expand his family's manufacturing business, and Kelly was busy with her MBA program. Within a few years, they were leading separate lives.

She tried to blame his lack of involvement on his busy schedule, but she soon noticed she felt alone even when he was home. Her marriage didn't resemble the contented companionship she had witnessed between her parents. It certainly wasn't anything like what she had dreamed of when she was young. She had imagined she and her husband playing tennis, taking photography classes, organizing community festivals, and touring antiques auctions to find treasures for their home. But they had even stopped going to the college football and baseball games they used to enjoy while they were dating. Her husband grew so distant that eventually he refused to accompany her to weddings, work banquets, and family gatherings. She was embarrassed to show up alone and was constantly making excuses for him.

She had been waiting for their lives to settle down before starting a family, but it had become clear they didn't have a strong enough foundation to survive the stress of having a child. Plus, Kelly had a greater fear: members of three generations of her husband's family had shown signs of mental illness. Her husband frequently wrestled with depression and then went on manic spending sprees that made her constantly nervous about their ability to pay off their credit card balances. Kelly didn't know whether his behavior resulted from a gene that could be passed on to her child, and she didn't want to take any chances. Still, every few years her husband would ask, "Why don't we think about starting a family?"

Kelly would try to buy time, saying, "Let's wait until we move" or "Let's wait until we finish the new building for the business." Over the years, he stopped bringing up the subject altogether. She occasionally felt longings to have a baby, but she didn't have any desire to have a baby with *him*. After eleven years of sticking out the marriage, she asked for a divorce.

Kelly never imagined she'd be single and childless at thirty-seven. But freeze her eggs? That sounded drastic. Her ob-gyn had never mentioned it. She didn't know anyone who had done it. She hadn't even heard about it until Kathy had brought it up.

When Kelly divorced at age thirty-five, she figured she'd remarry in time to have a baby. A couple months after her divorce, she was already in another serious relationship. She had met Steve, a tall, well-dressed management consultant, when they were both dropping off charity Christmas gifts at their church. She usually attended the later service and had never seen him before.

Kelly admired how different Steve was from her ex-husband and fell for him quickly. Not only was Steve affectionate and respectful of her feelings, but he wanted to be with her constantly. Several nights a week, they made dinner together; he worked as her sous-chef, chopping all the vegetables they had bought at the farmers' market. While they cooked, he would pour her a glass of Chardonnay and play his favorite CDs of Miles Davis or Thelonius Monk. Kelly loved how he wanted to know her opinions about everything: religion, work, philosophy. She had become so used to her ex-husband's silence that she had forgotten how good it felt to share simple conversation.

At age forty-four, Steve was nine years older than she, divorced, and the father of a teenage girl. Kelly let him know within the first couple of dates that she wanted children of her own. However, she waited four months into the relationship before asking his feelings on the topic. "Do you think you'd want more kids?" she asked him over sushi one weekend.

"I don't know," he confessed. "I didn't really enjoy it in the beginning." His marriage had ended when his daughter was three, and he had a hard time being a single dad. He admitted that he had even undergone a vasectomy several years earlier to avoid an accidental pregnancy. However, he stressed, he wasn't completely against the idea if she really wanted one. "I'd do it for you," he said. To prove he was serious, he emailed her some links on vasectomy reversal a few days later.

At first, Kelly was wildly flattered, but it soon became clear he was considering taking on such a huge responsibility only for her sake. She didn't want a "duty dad" for her baby; she wanted to be able to share the excitement with someone. She also wasn't sure Steve had natural nurturing instincts. He was a good father to his daughter, but she was bothered by his short fuse, especially the way he became easily frustrated while teaching her how to drive. Plus, he never made an effort to engage the children of her friends. Once when they walked by a playground filled with preschoolers, he commented, "I found it hard raising a kid that age." Kelly cringed and didn't bring up the subject again.

By their first anniversary, Kelly knew she didn't want to marry Steve, but she also didn't want to end the relationship. She loved being with him and was scared she'd never meet anyone else she liked better; there weren't a lot of available men in her town. She felt her only choices were to break up with him or accept that she'd never be a mother. She couldn't bear either option. So she let the relationship plod on, enjoying the present but never feeling hopeful about the future.

When Kelly turned thirty-seven, it seemed as if her whole office had suddenly become worried about her biological clock. She casually mentioned what Kathy had said about egg freezing to her coworker Courtney, who was thirty-two and pregnant with her first son. Courtney had the reputation of being a free spirit who said whatever was on her mind and would get to the heart of sensitive subjects that others would dance around, such as Kelly's love life. "Kelly, when are you going to dump that guy and find a husband?" she'd ask. Or she'd pop into Kelly's office and ask, "Did you call about egg freezing?"

"No," Kelly responded.

"Well, are you going to?" Courtney persisted.

"No," Kelly repeated exaggeratedly, as if she were a fourteen-year-old being harangued by her mother.

"Shouldn't you at least hear about it?" Courtney asked.

Kelly's answer was always "No." But when she heard a few weeks later that Dr. Kevin Winslow, a fertility specialist from the Jacksonville clinic Kathy had mentioned, was going to speak at a local hospital as part of its continuing medical education program, she promised Courtney she would go.

During the presentation, Kelly didn't know many of the medical terms, but her cheeks grew hot as she understood the message: at nearly thirty-eight, she was approaching the end of her childbearing years. "Women are born with an ovarian reserve of one to two million eggs, which die off steadily after birth," Dr. Winslow lectured, explaining that women lose more than a thousand egg follicles during every menstrual cycle as their body chooses one egg to mature and send down a fallopian tube to be fertilized. "By puberty, she has only 400,000 follicles left, and by menopause, her reserve has been exhausted." Not only did women Kelly's age have fewer eggs, he said, but the eggs that were left had a greater risk of being chromosomally abnormal. In fact, Winslow said, modern fertility medicine could overcome nearly every other problem in making babies except old eggs.

Kelly's head grew fuzzy with anxiety.

Then Winslow discussed a technique called oocyte cryopreservation that stopped the eggs from aging. "For women who feel strongly about having a genetic child, freezing their eggs is a valuable option," he concluded.

Kelly introduced herself afterward and mentioned that she might be interested in learning more about egg freezing.

"How old are you?" Winslow asked.

"I'm thirty-seven but turning thirty-eight soon," she replied.

Winslow, whose height and directness gave him a commanding presence, explained that his cutoff was a woman's thirty-ninth birthday. He didn't believe eggs older than that were worth saving. "You need to come see me soon," he urged.

Kelly loved the idea of egg freezing, but Winslow charged $10,000. She had no idea how she could come up with that kind of money.

A few days after Kelly turned thirty-eight, she found a note in Courtney's handwriting taped to her office phone. "Call Dr. Winslow TODAY," it read.

An hour later, she picked up the phone.

Hannah

"This can't be happening again," Hannah Burke frantically thought to herself. "Not here! Not now!" She had just started a much-anticipated ski vacation with Marcus, her boyfriend of four months. After driving nearly five hours from Seattle to Whistler, they were finally drinking cold beer in the outdoor hot tub of their hotel. Hannah tried to focus on the view of the snowy mountains, the steam coming off the water, and the snowflakes catching in her hair. But she couldn't shake off the unmistakable symptoms—racing heart, shortness of breath, hot flashes, tingly fingers—that meant an embarrassing panic attack was on its way. She knew that within minutes, she would be crying uncontrollably and starting to hyperventilate.

"I've got to go back to the room," she said as she abruptly climbed off his lap and grabbed her towel. She barely had the key in the door when the tears came. A few minutes later she called her mom, who reassured her that she hadn't ruined the trip. Feeling calmer, Hannah wiped her face, washed off her smeared mascara, took a few swigs of wine to soothe her nerves, and put on her best face for Marcus. Thankfully, he acted as if nothing was wrong.

Hannah, age thirty-seven, had experienced a few minor panic attacks in her mid-twenties when she was attending fashion design school in Manhattan. But they had disappeared for more than a decade, only to resurface a few weeks after she met Marcus on a blind date set up by a friend. She tried to control the attacks by talking to a therapist and taking the anti-anxiety drug Xanax, but the episodes were still frequent and frightening. "Why now?" she asked herself. She was in the most promising relationship she had known in years, and she was sure this trip would determine whether they became a solid couple. Marcus, an architect who was a year younger than she and ran his own business, was everything she wanted: good-looking, creative, funny, and successful. "Oh my God! He is the one!" she told herself after their first date. The *one* meant her future husband.

Throughout her life, she had expected marriage would just happen for her. Friends said she was a catch: a beautiful girl with a compact curvy body, chic brunette bob, and big smile. She had an enviable job as a men's clothing designer, owned her home, and maintained a solid group of friends from high school and college. She was a nice person, and nice people got married. But at nearly forty, she hadn't even come close.

She had never kept a boyfriend for more than a couple of months. In her twenties, she had a reputation as a party girl who enjoyed making out with men she met at bars or parties, never interested in anything serious. During her thirties, she pursued longer relationships, but they never went anywhere. She usually became attached quickly, especially if she slept with someone, and then became anxious about whether he liked her back. She tried to project a strong and confident image to make herself more attractive, but she couldn't maintain it for long. If she sensed a man was losing interest, she dug in deeper by trying to become the kind of girlfriend she thought he wanted. If he liked the outdoors, she would take up mountain biking. If he liked sushi for dinner or sex in the morning, she was game. But the more she accommodated men, the more they withdrew. Instead of getting reassurance that she was desired, she got dumped.

The cycle had become so painful that she avoided watching romantic comedies, such as *Kate and Leopold* and *Notting Hill*. She was never inspired by the storylines about characters falling in love; she just became depressed because she feared she would never experience what they did.

Hannah had enjoyed her single days, but for years she had been ready to move on to the next phase of her life. When she bought a condo at age thirty, she began to think about financial investing and wished she could make long-term plans with a partner. She loved the idea of building something with someone. She also admired her married friends' lives and the way they cooked dinner together, talked about their upcoming trips, and shared parenting duties and inside jokes. She craved that kind of companionship.

She was used to romantic letdowns, but they had become more excruciating ever since her ob-gyn had explained that her fertility was declining. "I notice that you're thirty-five," the doctor had said during Hannah's annual exam. "I just want to ask you about your plans to have a family. You look great, but your eggs are getting older." Now the risk of a failed relationship wasn't more loneliness; it was the risk of not becoming a mother.

In contrast to today's increased media coverage of women's fertility, many women in Hannah's generation had little knowledge about their biological clock. And doctors are divided about whether it's their responsibility to broach the subject. While some ob-gyns believe it is, others prefer that patients initiate the conversation to avoid making women who want children anxious or making those who don't want children uncomfortable. Hannah's ob-gyn decided to walk the fine line between educating and upsetting her patient.

Hannah was indeed shocked when she learned that many women who became pregnant in their late thirties or early forties had resorted to numerous rounds of IVF treatment or donor eggs. She had thought she still had plenty of time. She was even more dismayed when her doctor told her about her available options: If she didn't have a partner, she could visit a sperm bank and have a baby on her own. Or, if she wasn't ready to have a baby now, she could have her eggs extracted, fertilized with anonymous sperm, and frozen as embryos, which could be transferred to her uterus later. That would give her time to find a husband, but the baby wouldn't biologically be his. Hannah quickly dismissed these options. She didn't want to be a single mom. Even though she worked full time, she didn't make enough money to hire a nanny. And at thirty-five, she wasn't ready to give up her dream of finding a husband. She wanted to have a child with him, not a stranger.

Hannah often recalled what a fortune-teller had told her on a work trip to Hong Kong: that she would know great joy and love in her lifetime. Except the fortune-teller had left out an important detail: When was great joy and love supposed to show up? In time for her to have children? Or was Hannah supposed to cut her losses and just focus on finding someone to love her?

Hannah knew intellectually that she would like to have children, but she didn't allow herself to want them. She tried to ignore the topic altogether: she didn't bring it up on dates; she didn't coo at infants or admire baby clothes; she didn't talk about baby names with her friends. But the strategy didn't work very well. Babies were still on her mind, and the pressure to make her relationship with Marcus work felt unbearable.

Hannah was both thrilled and terrified when he invited her to Whistler over Thanksgiving weekend. Some of his friends were also going, and she knew this was her chance to make a good impression. But she had never traveled with a boyfriend before and was seized with self-doubt: What if she wasn't funny or sexy enough? What if her butt looked too big in her bikini? What if she wasn't a good enough skier? When should she let him pay, and when should she offer to contribute? Where should she use the bathroom?

Then there were the panic attacks. How would she explain those? She hadn't experienced them with other boyfriends, and she couldn't imagine what Marcus would think if he witnessed her coming unglued. Her therapist had advised that she warn him, just in case. So on the way to Whistler, Hannah took a deep breath and recited her therapist's suggested script. "I haven't traveled much with boyfriends, and I'm a little nervous about going on this trip," she said. "I want you to know I'm dealing with some anxiety issues." She explained that if she had a panic attack, she might need some time to be alone or write in her journal. Or she might take the medicine she had brought to calm her down. "I just want you to know these aren't about you at all," she explained. "I'm just a nervous person."

She was relieved when Marcus seemed sympathetic. Still, she wondered if he privately thought she was a nut job. She made it through the rest of the trip by taking Xanax and drinking beer at lunch or Cape Cod cocktails *après-ski*. She tried to be extra passionate during sex, but Marcus felt distant, which made her even more nervous.

She wasn't so sure her panic attacks *weren't* really about him. She wondered if there was something about Marcus that brought out her worst insecurities. In the beginning, he had wanted to see her nearly every night of the week. He often invited her over to his house and cooked her dinner, including some specialty from his home state of Texas involving lots of nacho cheese sauce. But when she left his house in the morning, she didn't feel a lingering contentment after spending the night with a new lover. She agonized over whether she had made a good impression.

Several months into their relationship, Marcus started keeping her at arm's length. He stayed in town for Christmas but opted to spend the holiday alone rather than with her family. "I'm not much of a holiday person," he explained. He declined to meet any of her friends. When he mentioned that his mom was planning to visit, Hannah piped in, "I can't wait to meet her."

"I'm not sure that's going to happen," he responded.

Hannah couldn't stand the uncertainty of not knowing whether this guy cared about her. She tried to remind herself that he liked *something* about her, since they had been together for six months. Then she tried harder to make it work. If Marcus wanted to see her, she invited him over. But if he said he was tired, she offered to drive to his house. If he wanted to wake up at the crack of dawn to be the first on the mountain during ski trips, she wrenched herself from bed rather than insist on sleeping in.

The night she returned from an exhausting weeklong business trip in Germany, Marcus said he needed to talk to her and was on his way over. She knew from his tone that it wasn't good news. Within minutes of arriving at her house, he announced, "This isn't working out. I don't want to pursue a relationship with you." She was calm during the breakup, but as soon as he left, she dissolved into tears. The crying quickly escalated into deep wracking sobs. She had suffered from breakups before, but they had never hit her so hard. Whatever flimsy sense of self she had possessed before now seemed nonexistent. She couldn't summon any perspective to tell herself there would be others. She was certain she had lost her last chance to get married—and her only chance to have a family.

In a haze of jet lag and rejection, she felt more alone than she ever had before.

Sarah

I estimate that I went on dates with at least fifty men before I met NYCPaul. Over countless glasses of wine in dark lounges, I identified common interests, monitored my body language, twirled my hair, did breathing exercises beforehand to appear relaxed—all in the hope of creating the right conditions for a spark to ignite.

Then, with NYCPaul, some curious mix of attraction, banter, and love of happy hours just worked. We spent hours talking about our past relationships and all the places in the world we wanted to visit. We dissected plotlines to the HBO show *Six Feet Under*. We went swimming and invented kissing games in between laps. Within a few months, I thought the musky scent at the back of his neck was the best thing I had ever smelled.

I had never been so happy. Paul surprised me with dinner reservations at restaurants with blue backlit bars and ceviche on the menu. He snuggled up next to me at movies, clipped story ideas for me, brought me coffee in bed, and made a big fuss over whatever buckwheat pancake blob I served him for brunch. We didn't argue; we simply discussed conflicts and arrived at reasonable compromises. He laughed at my jokes and asked me to repeat them. Whenever I cooked, he popped into the kitchen to rub my shoulders.

Paul thought the things I did that annoyed everyone else—dragging too much crap through the airport and expecting him to help carry it; repeatedly changing restaurant tables because I could feel a draft; speaking in single-word sentences before my morning coffee—were cute. "I love all your little ways," he'd tell me. No one had ever loved all my little ways, not even my mother. When we walked down the street, he puffed out his chest, as if to show men who glanced at me how lucky he was. At thirty-five, I had never felt so adored. I had never felt so beautiful. I was starring in my own romantic comedy, and it was one I hadn't had the imagination to write myself.

Still, I wondered what he meant by answering "Maybe" to the question on his online dating profile about whether he wanted children. Did he really want them but was leaving the door open in case he met a woman who already had them or *couldn't* have them? Did he really not want kids, but didn't want to sound like a jerk by marking "No"? Or could he really go either way?

I didn't ask at first. I figured a man who was dead-set against children wouldn't pursue a woman who was clear she wanted them. Besides, I was enjoying my movie too much. But within a couple of months, I couldn't ignore the topic anymore. So after brunch, I coolly asked, "So what's up with that 'Maybe' on your dating profile?"

After a long pause, he responded, "Well, I don't want kids right now. But who knows about the future? I guess I'm open." As for now, he asked if I could imagine a life in which two people just focused on each other. We could jet off to Miami, where he had invested in a condo, or go to the movies whenever we wanted. He trotted out studies claiming that people without children were happier and that their marriages were stronger. Certainly they had more sex.

I'd read the same reports and had been noticing a glut of studies concluding that having kids was a really bad idea. A recent eight-year University of Denver study of 218 couples found 90 percent experienced a decrease in marital happiness once the first child was born. Other studies claimed that satisfaction didn't bounce back until after the last child left home—*if* you didn't succumb to divorce in the meantime. Even if your marriage fared better, you could never count on it to be the same. The study didn't mention how much it cost to raise a child: at least $250,000 in 2009. (It's since gone up.) Who wanted to sign up for financial sacrifice, drudgery, divorce, and, if he still stuck around, a lousy, stressful, power struggle of a marriage?

Kids wrecked overall happiness too. A study of thirteen thousand people found that parents were more depressed than people without kids. An article on happiness myths that was published in a Harvard Medical School consumer publication argued that children did not necessarily bring bliss. In fact the article referred to a survey of mothers who were asked to rank their happiness throughout the day. It turns out that they were more content eating, exercising, shopping, napping, or watching TV than spending time with their kids.

The mounting evidence, including a burgeoning genre of "keeping it real" mommy blogs and parenting tell-all books, didn't even begin to describe what pregnancy and birth did to your body: weight gain, stretch marks, no sex for six weeks after delivery, the reassignment of your breasts as milk producers and spit-up rags. The clincher was a *New York* magazine article about a popular parenting website called UrbanBaby.com in which lonely, tired, and bored moms revealed on message boards how much they hated their lives and husbands, who were often cheaters or workaholics. The women complained about taking on the grunt work of child rearing and were resentful that their husbands refused to help out, pestered them for sex, or wouldn't put out at all. The women drank too much, wished they hadn't given up their careers, or, if they worked, wished they could quit and stay home. Many predicted their marriage was doomed.

I usually tried to dismiss such bitching as that of vocal individuals and not reflective of a group as a whole. I was convinced the studies were inaccurate or asked loaded questions. Then the Rutgers Marriage Institute published "Life without Children," an essay theorizing that since adults spend most of their lives without small children, the idea of giving up ownership of your time leads to a phenomenon known as "mommy shock." I nearly yelped when I read this sentence: "Increasingly, Americans see the years spent in active child rearing as a grueling experience, imposing financial burdens, onerous responsibilities, emotional stress, and strains on marital happiness."

I didn't want grueling. I had just escaped grueling.

I managed to comfort myself with an often overlooked fact in the "children as misery" literature: people kept having them. Most of my friends had more than one child; apparently they enjoyed the drudgery. When I looked deeper into the studies, I found a lot of nuance not reflected in the headlines, such as the finding by Scott Stanley, a researcher at the University of Denver, that couples without kids also experience decreased marital satisfaction over time, although at a slower rate. There was also some hopeful news: couples with longer marriages or higher incomes survived the shock much better, and some couples reported that their marriage even grew stronger after they had children. Most important, even couples whose marriage was rocked by kids felt good about building a family, a reward Stanley called "family happiness." And according to a Pew poll, while only 40 percent of parents with kids under eighteen believe children are important to a successful marriage, a whopping 85 percent list their relationships with their kids as a top contributor to personal fulfillment. Relationships with spouses receive similar high marks. Less than a third of respondents think careers or free time significantly contributes to their happiness.

But no carefully crafted argument or published study could compete with my family fantasies that Paul had unknowingly unleashed. Within a few short months, Claire got promoted from abstract Saturday sidekick to a fleshed out five-year-old. The three of us made Mickey Mouse pancakes, sledded in Central Park, and dressed up the dog in funny costumes. Paul taught her how to play chess and quizzed her on moves as they walked to school. Sometimes I imagined him sheltering her baby brother from the snow under a classic navy wool pea coat, a leftover Gap-inspired daydream from college. After dinner we made up songs and told stories during bath time. When the kids were in bed, Paul and I shared a bottle of wine and recounted our day. It was a lovely life.

It wasn't lost on me that I was letting my imagination run wild about a man who wasn't sure he wanted to be a father. But the deeper I fell in love with Paul, the more I wanted to create a family with him. I knew he would treat our children as well as he treated me, and my heart swelled when he made my friend's baby laugh at his silly voices or bragged about teaching kids in his building how to swim. For the first time in my life, I had an outlet for all my baby longings.

My maternal pangs were in full swing. Tears streamed down my cheeks when I heard the joy in my brother's voice after my niece, Kate, was born. I grew jealous when I watched my mother decorate the family bassinet in white tulle and pink satin ribbons. My family competed to hold her.

I was also no match for Facebook, where I was deluged with pictures of my friends enjoying their children. They posted snapshots of themselves feeding babies, bathing babies, pushing babies on swings, and hoisting them in the air à la *The Lion King*. Their online photo albums were filled with babies dressed in floppy hats or Halloween costumes, plopped on pumpkins, picking strawberries, toddling down the beach, digging into birthday cakes, snuggling with their dads, playing with their grandparents, and sitting in every swing or Bumbo Baby Sitter ever sold at Babies R Us. Facebook wasn't the place to complain about how little sleep or sex they were getting. This was Boast Central.

Sometimes I felt overcome with sadness as I peered into their lives. "Someday you'll have all these things too," I told myself in my usual refrain.

But when was *someday* supposed to happen?

On the afternoon of my thirty-sixth birthday, I parked myself in a stifling Starbucks and furiously scrawled out my new fertility math on the back of a business card I found in my wallet.

Paul had said he didn't want kids now but was undecided about later. *If* he could agree to start a family within a couple of years, I could have my first baby just before I turned thirty-nine and quickly have a second. It wasn't ideal, but I thought it could work, assuming I had no problem getting pregnant.

Though I hadn't yet said anything to him, apparently I wasn't difficult to read. I was only two days into a family reunion in Lake Tahoe when he called from New York. Random relatives were already grilling me about that "nice young man" they'd heard I'd met.

"We've been dating almost a year," he said.

"Yes," I responded hesitantly. I froze in the hotel hallway.

"I'm very happy and love you deeply. But I'm guessing that you want to be married and pregnant by a certain age," he said. "I just need to let you know that I'm not on a timeline."

"I'm not sure what you mean," I replied, growing nauseous. Over the next two hours, we parsed out the headline. He didn't want to close the door to *ever* having children, but he also felt a responsibility to be honest. Being a dad wasn't a priority. If I felt I had to have a baby within a certain time frame, he didn't want to stand in my way. I reassured him that I didn't just want a baby. I wanted him, and I was willing to see where our relationship could go *if* he thought it was going somewhere.

He said that he did.

Afterward I walked around in a daze. Was he just scared and openly sharing his doubts, as anyone should be able to do in a good relationship? Or had I experienced one of those seminal moments, such as the ones I'd heard about from friends and Dr. Phil that usually involved *this girl* who didn't listen when her boyfriend said that he didn't want to marry her and have a family? The chorus would promptly sing, "Oh, how could she be so dense? She's in denial. She's hearing what she wants to hear. She should stop wasting her time!"

My stepfather had scored tickets to Crosby, Stills, Nash and Young, and that evening the whole family went to the concert. I purposefully sat on the other side of my brother so that my mom couldn't tell I was upset. I tried to act as if nothing was wrong, but when the band started singing "Our House," I couldn't stop the tears. My brother noticed and handed me a beer.

I loved that song and used to play it constantly in college. It's about making your home. My new straight-talking therapist liked to say, "You're sitting on your home. It's your ass." In other words, you have to be content wherever you were. But home as I remembered it from my childhood was about babies in high chairs at the dinner table, every member in his or her place, eating zucchini we had picked from the garden. I felt as if my whole adult life had been about finding a way to get back there.

I ached to make a home with Paul. I loved waking up with our bodies pressed together. I loved talking to him from the shower while he shaved in the mornings. I loved the smell of rosemary chicken roasting on Sunday afternoons. I loved a bed messy with newspapers. I loved planning vacations, knowing the names of his cousins, and discovering favorite HBO shows together. I craved creating layers of belonging that I envisioned would keep me warm and secure. I just wanted Claire to be invited too.

But at my parents' San Diego home, instead of vegetables, I got lectures. "Why did you become involved with a man who is unsure about having kids if that is truly important to you?" my mother implored. "You don't have any more time. Paul is forty-two! He should know by now."

My therapist was just as blunt. "Maybe you really don't want kids," he said, baiting me. "If you really wanted them, you would have found a way to have them by now." He explained that I had to pay attention to my walk, not my talk. His advice: Make some difficult decisions, or convince Paul that having kids with me would be a good experience.

I definitely wasn't ready to let Paul go. When we read in bed at night, he'd smile at me and say, "Do you know how loved you are?" I did, and I couldn't get enough of him. He brought out the best in me. I worked better. I slept better. I exercised more. I was nicer. I even felt lighter—except for my thirty-six-year-old ovaries, which suddenly felt like casaba melons of responsibility.

"I wish I could just get these eggs out and freeze them," I complained to my friend Janelle one night over drinks after attending an alumni happy hour.

I half-expected her to say, "Freeze your eggs? That's crazy. You're only thirty-six!" Instead she said, "That may not be a bad idea. Actually, I've heard the success rates have gotten better."

A couple months after the Tahoe talk, I asked my ob-gyn about egg freezing. She confirmed the promising reports and thought it might be worth looking into if I didn't foresee having kids in the near future.

I knew little about what egg freezing entailed, but on the subway ride home, I pulled out the business card on which I had written my future timeline and added a new line item: "Age 36: Freeze eggs."

2. *Hypotheses*

Dr. Michael Tucker

In the spring of 1986, the British embryologist Michael Tucker sat in his office at an IVF clinic in Hong Kong, staring in disbelief at the British medical journal *The Lancet*. The latest edition contained an article with the perfunctory title "Pregnancy after Human Oocyte Cryopreservation," announcing that a fertility doctor in Australia had achieved a twin pregnancy after freezing and thawing his patient's eggs. Tucker was surprised someone had finally solved what had been a medical mystery for years. Although scientists had been able to freeze and thaw sperm since the 1950s and embryos since the early 1980s, eggs were thought to be impossible to preserve.

Because eggs (oocytes is the scientific term) are the biggest cells in the body and contain mostly water, the freezing process creates destructive ice crystals. In a normal egg, twenty-three chromosomes are neatly positioned in a straight line on a meiotic spindle. In previous trials with frozen and thawed eggs, however, these pieces of DNA often ended up clumped together or scattered all over the place. If a thawed egg was fertilized, the two pairs of chromosomes might not combine properly, and some of the egg's chromosomes would be missing. Sometimes the spindle completely disappeared from view.

In the article, Dr. Christopher Chen from the University of South Australia in Adelaide wrote that he had frozen forty of his patients' eggs and that 80 percent had survived thawing intact. The most important news, however, was that the majority of the good eggs appeared to fertilize, and more than half developed into viable embryos. Dr. Chen transferred two to three embryos into each of four women's uteruses, and one became pregnant with twins.

Tucker, twenty-nine, who had left a career in the science of breeding cattle and sheep three years earlier, had been looking for a way to make his mark in this new field but wasn't sure what to make of Chen's research. *The Lancet* article was light on details, and Tucker questioned why the doctor had frozen his research eggs for only five hours and had transferred the resulting embryos into just a handful of patients instead of conducting a bigger study. And since Chen was the sole author of the article, he lacked the support of fellow researchers.

In fact some of Chen's colleagues wondered whether his pregnancies were really from frozen eggs. Chen had given his patient fertility drugs and surgically removed six eggs through an incision in her abdomen, potentially leaving some behind that could have been inseminated naturally. It would be a few years before doctors started using the more accurate transvaginal ultrasound, in which they insert a probe into a woman's vagina and see pictures of her reproductive organs via sound waves on a monitor.

The news of the births of David Clive and Cheryl Catherine Chrischen (named after the doctor) to a twenty-nine-year-old woman was greeted quietly by the world press, relegated to another "first" in a fast-moving field with unimagined applications and consequences. The science seemed so routine and easy. Local media reported that Chen had frozen only three of his star patient's eggs, and two had become live, healthy babies.

Even if Tucker had enough information to repeat Chen's study in his home lab, he wasn't sure how useful it would be to his IVF patients. He was intrigued by the scientific challenge, but he didn't understand why a woman would want to freeze her eggs. The patients he saw wanted to be pregnant as soon as possible. Tucker concluded that egg freezing wasn't worth his time.

Shortly thereafter Chen announced a pregnancy for a thirty-seven-year-old whose eggs were thawed after being frozen for four months. However, Chen never reported another one, and his frozen-egg successes were forever regarded as impressive flukes. Over the next couple of years, two European teams also announced one birth and two pregnancies (which ended in miscarriage) using frozen eggs, bringing the worldwide total to five. Yet the poorly documented cases seemed to be lucky breaks as well: those results were never replicated by the doctors or anyone else.

If that wasn't enough to discourage scientists from taking up egg freezing, Australian researchers published a paper in 1988 documenting that only 5 percent of embryos made from frozen mouse eggs became viable fetuses. Since mouse eggs were readily available and cheap, they were frequently used as litmus tests for human eggs. If mouse eggs couldn't survive, it was assumed that human eggs wouldn't either. Even more disturbing was the fact that such embryos were three times as likely to be chromosomally abnormal as those made from fresh eggs.

Few scientists were willing to risk using frozen human eggs. Louise Brown, the first IVF baby made from *fresh* eggs, was only ten years old, and the reproductive medicine world was crossing its collective fingers that she and subsequent IVF babies wouldn't develop long-term health problems. It didn't make sense to push the boundaries with eggs that had been potentially damaged from freezing.

Meanwhile Tucker had focused on another medical challenge: how to improve fertilization rates with poor quality sperm, which account for some 40 percent of infertility problems. He published several articles about a now-outdated procedure called gamete intrafallopian transfer (GIFT), in which sperm and eggs are mixed together in a catheter and surgically placed inside a woman's fallopian tube to create an embryo that travels to her uterus. GIFT became a popular option for religious patients opposed to creating embryos outside the body, but it was also a solution for men with low sperm count or "weak swimmers" that might not find their way from the vagina to the tubes on their own.

However, the procedure's success still depended on the sperm's ability to penetrate the egg. Even traditional IVF, in which scientists let millions of sperm fight over an egg in a Petri dish and then transfer the resulting embryo directly into a woman's uterus, requires a sperm strong enough to burrow into the egg.

While in Hong Kong, and later in Atlanta, where he moved in 1989 after the Tiananmen Square massacre drove out the foreigners who were his clients, Tucker and his colleagues experimented with micromanipula tion, a technique using joysticks or mechanical hands attached to a microscope, to help such slacker sperm make it into an egg. In one method, they cut open an egg, but too many sperm got in. (An egg fertilized by multiple sperm might develop into an abnormal embryo that won't survive.) In another, they used a pipette to insert a sperm just under the egg's outer membrane in hopes of giving it a head start, but a lazy sperm still had trouble penetrating through the inner protein coat. Finally, after designing a needle that didn't damage the egg, they developed the technique of injecting a sperm directly into the center of the egg. That was the jackpot.

But Tucker missed the glory of being the first to discover it. In the summer of 1992, a Belgian team announced the first births in the world of four babies conceived via intracytoplasmic sperm injection (ICSI). The fact that Tucker's clinic, Reproductive Biology Associates, had several such pregnancies under way at the same time was no longer big news.

Scientists had finally figured out how to make a baby in the most direct way possible, and ICSI is now a standard procedure at most fertility clinics. But there's reason to be concerned about the long-term effects. In nature, these weak sperm weren't meant to create babies. Is ICSI propagating genes responsible for male infertility or other medical problems? That answer won't be known for years, considering that the first wave of boys born from ICSI is now approaching reproductive age.

Tucker and his colleagues couldn't have known it at the time, but ICSI would prove to be the missing link that made the fertilization of frozen eggs possible. The reason: thawed eggs are hard to penetrate.

But that discovery would come later. At this point, scientists still didn't know how to freeze, thaw, and fertilize eggs without risking the creation of a DNA disaster. One young Australian embryologist named Debra Gook was convinced she could solve the puzzle. In 1994 she generated buzz at a conference of the Pacific Coast Fertility Society in southern California with her findings that eggs could be frozen and thawed without disrupting their chromosomal structure. Gook, who became interested in egg freezing after watching her best friend lose her fertility during cancer treatments, had compelling evidence that human eggs could make it through the egg-freezing process intact. The trick, Gook explained, was to use the chemical compound propanediol, which slowly leeched the water out of the egg before freezing to prevent ice crystals. Her team at the Royal Women's Hospital in Melbourne froze approximately 170 human eggs, and nearly two-thirds survived thawing. The outer layer, called the zona pellucida, wasn't damaged, and the eggs retained their shape. Most important, about 60 percent of those had a normal barrel-shaped spindle with its chromosomes in a reassuring line. A damaged spindle looks misshapen, like a star, elongated triangle, or cockleshell; its chromosomes appear fuzzy, overlapping, or dispersed. In presenting her findings, Gook dismissed the alarming Australian study with mouse eggs that had been published five years earlier. She claimed it wasn't a fair comparison, since the spindle in human eggs isn't as fragile.

Tucker happened to be at the same conference giving a lecture on ICSI. Impressed by Gook's work, he asked how the thawed eggs were fertilizing. Gook explained that the strict laws in her home state of Victoria prevented her from experimenting with embryos, but that she was allowed to fertilize the eggs as long as she destroyed them within twenty hours of insemination. She told Tucker that she was working on another study and had found that nearly half of the thawed eggs fertilized. She believed the number would have been higher, except that thawed eggs presented an unexpected challenge. Freezing appeared to harden the zona, which made it difficult for the sperm to break into the egg.

"You know, with ICSI, you might get better fertilization rates," Tucker said, explaining how the technique would bypass the hardened zona by manually inserting the sperm into the center of the egg.

Gook said she and a colleague had already been trying it out at the University of California in Irvine. Two years later, Gook and her team published another study showing that thawed eggs fertilized by ICSI consistently developed into robust embryos.

The next challenge was to see if one would grow into a baby.

• • •

More than a decade after learning of Chen's frozen-egg babies, Tucker was intrigued by the scientific challenge of tackling one of the last major hurdles left in the field of reproductive science. But he had stopped short of investing too much time in the pursuit. He wasn't certain how the procedure would benefit his patients. Over the years, he and his colleagues had theorized that some women *might* be interested in freezing their eggs to delay having a baby, in case they hadn't yet found partners or needed time to focus on their career. But frankly, he didn't want to offer the service to them. Whenever he mentioned at dinner parties that he was researching egg freezing, women jumped all over the topic. "Oh, my sister could use that. She's forty," they might say. Some would ask directly, "How can I do it?" Female colleagues would corner him at conferences, saying, "I spent all these years at medical school and building my practice. Now I'm running out of time. Will you do me a favor?" The intense interest unnerved him.

Tucker thought that many Clock Tickers were used to being successful in most areas of their lives and were rankled that fertility was out of their control. He didn't like the idea that these women were using egg freezing as an excuse to put off having a baby and would rely on him to perform magic later. How could he possibly meet those expectations? He didn't want to be the bearer of bad news when their eggs didn't work later. He told the Clock Tickers no.

Besides, if he wanted to explore egg freezing seriously, he needed a quick turnaround on research results, and women who had frozen their eggs in hopes of pausing their biological clocks wouldn't likely use those eggs for several years.

Still, by the mid-1990s Tucker had become inspired to try his hand at egg freezing when he found a reason to offer the procedure at his Atlanta clinic: he could freeze the eggs of women who were willing to donate them to infertile women. It would be cheaper and more efficient to offer patients frozen donor eggs than fresh donor eggs, which usually cost at least $15,000 a cycle to cover donors' compensation fees and medical costs.

The first successful pregnancies from donor eggs were reported in 1983, and the procedure's popularity had grown so quickly that twelve years later, at least 10 percent of Tucker's patients requested them. A woman with diminished egg quality or who has undergone early menopause can have another woman's eggs fertilized with her partner's sperm and carry the baby as if it were her own genetic material. Since the recipient can experience giving birth and nursing the baby, many women consider donor eggs an attractive alternative to adoption. They also don't have to worry about custody issues or whether the biological mother used drugs or alcohol during the pregnancy.

Ordinarily patients purchased one batch of a donor's eggs; all were fertilized, and any extra embryos were frozen. Several patients could go in on the cycle, but that arrangement required a major logistical feat: doctors had to sync up donors' and recipients' ovulation cycles, and it was impossible to predict how many eggs a donor might produce. If the harvest was enough for only one recipient, the others who were sharing the cycle were left in the cold.

But if doctors could freeze all of a donor's eggs, they could distribute a few at a time to many recipients whenever they needed them. For example, a donor's harvest of twenty eggs could potentially help as many as five patients, who might buy four eggs apiece.

Tucker wanted to set up a bank for donor eggs, since IVF patients likely would use them in the near future and give him plenty of opportunities to research and refine the technology. In late 1994 he got the go-ahead from his partners at his clinic to pay twenty-one women in their twenties and early thirties for their eggs. In one of his first experiments, only a quarter of three hundred eggs survived freezing and thawing. But half of those were successfully fertilized through ICSI. He transferred several of the embryos to five donor patients and was thrilled when three women became pregnant—only to be deflated several weeks later when all miscarried.

Then came more bad news. Tucker learned about the work of a Korean group who had found surprisingly high rates of chromosomal and spindle abnormalities after freezing and thawing immature eggs (eggs that have yet to be matured with stimulation hormones). Although the Koreans were using a new fast-freezing technique called vitrification, which was different from Gook's slow-freezing method, the research was disturbing enough to prompt Tucker's partners to suspend the study. Egg freezing still appeared to be too risky.

But Tucker still believed in his frozen eggs and prepared a compelling case to convince his partners to reconsider their position. He argued that his team had frozen mature eggs, which were less fragile than the immature eggs used in the Korean study. He pointed out that the freezing formula was similar to a technique they used for freezing embryos, which had a long and successful history. He was reassured by his own experiments, which showed that embryos from frozen eggs developed normally, and insisted that his approach was far from reckless. He would transfer only a perfect-looking embryo and monitor the baby's development.

A year later Tucker's partners let him continue the study. After being approved by a hospital ethics review board, he offered the service free to needy patients who were willing to try frozen donor eggs. His team transferred embryos made from eggs that had been frozen an average of two years to fifteen women. Four became pregnant. Three miscarried. But a thirty-nine-year-old mother of two teen girls who had gone through early menopause was carrying twins. Tucker and his staff cheered when she made it past the first trimester, out of the danger zone for miscarriage.

Tucker wanted to revel in their success, but his nerves wouldn't let him. As a scientist who had worked on many cutting-edge technologies, he had lived through too much disappointment in the lab. He felt even more cautious because his patient was carrying twins, which was a risky pregnancy, especially for an older mother. He was relieved when an ultrasound showed the sixteen-week-old babies were developing normally.

Tucker thought he and his team would be the first to produce a healthy baby by using frozen eggs that had been injected with sperm. But they were too late again. One May morning in 1997, his coworker stopped by his office and asked, "Did you hear?" A group of doctors from

Bologna, Italy, had just announced the birth of a baby girl conceived from frozen eggs.

Tucker was crushed. Three months later his patient delivered twin boys, each with ten toes, ten fingers, and forty-six perfect chromosomes. (Although the Italians could claim the first birth in the world from a patient's frozen eggs conceived via ICSI, Tucker still won the honor of the first births from frozen *donor* eggs conceived via ICSI.)

The American media went nuts. Gina Kolata of the *New York Times* wrote that egg freezing would "make menopause obsolete."

Tucker, a tall, engaging man with a shaved head, made the rounds on most of the major broadcast networks. He tried to stay on message about the potential of egg freezing to help IVF patients, but nearly every media outlet mentioned how it could help women who were worried about losing their fertility before they were ready to have children. He was quoted in *Time* as saying, "We have leveled the reproductive playing field for women. We've allowed them to turn off their biological clocks."

Within hours of his first television appearance, his clinic was flooded with phone calls. Weeks later Tucker still had dozens of calls to return at the end of nearly every day. But in spite of his fear that "hordes of 25-year-olds" scared of being "left on the shelf" would swarm fertility clinics to freeze their eggs, as he told *People*, the vast majority of callers were anxious women in their late thirties and early forties.

Tucker felt an obligation to call these women back. He dreaded delivering the news that he didn't have enough clinical data to offer egg freezing as a routine service to patients. After inquiring about their age, he'd gently suggest they get blood tests to learn their current hormone levels as well as vaginal ultrasounds to see how many follicles they produced every month. The data could give them a sense of how much time they had left. Then he would try to console them by telling them they still had options. "Don't give up," he urged. "Even if you hit menopause, you can use donor eggs. We help women your age and older daily."

Tucker wasn't just reluctant to make tenuous promises to Clock Tickers. He had reason to be cautious. His research results were all over the place. In one batch of eggs, 50 to 60 percent would survive freezing and thawing intact, while in another only 5 to 10 percent would make it In his largest trial of more than three hundred donor eggs, about a quarter survived. He defended the disparity to dismissive colleagues by pointing out that he had not exclusively chosen the best eggs, which might have improved survival rates. He had also wanted to see how freezing affected older and immature eggs as well as eggs with strange shapes. Still, he wondered how he could sell a service based on those results. With embryos, on the other hand, he had a consistent 75 percent survival rate. Meanwhile medical supply companies were calling him to patent the freezing solution. He was being invited to speak at major conferences about egg freezing. But instead of enjoying the recognition, he had no confidence he could replicate his success. He was weary of the whole topic.

Besides, there were bigger problems to worry about. The expanding field of reproductive technology was plagued by births of multiples, low success rates, little federal oversight, overcharging, embryo custody battles, parental rights debates, lack of donor sperm screening, and even cases of lab mix-ups in which embryos were lost or babies were made with the wrong sperm, a problem that became most visible when white women with white husbands gave birth to mixed-race babies. Tucker couldn't indulge in pet research projects when he had to focus on improving record keeping, tightening the chain of custody in his labs, and enforcing new safety checks to make sure the right eggs were matched with the right sperm.

At the end of 1998, Tucker left Reproductive Biology Associates to help start another clinic. He left his egg-freezing research—and stockpile of frozen eggs—behind.

Sarah

When I told my editor at *Marie Claire* in the summer of 2006 that I was thinking about freezing my eggs, she thought other women might be interested in the topic and assigned me a story. I looked up egg freezing on the Internet and learned that a nonprofit women's group called Step Up was hosting a seminar to promote Extend Fertility, a referral service that paired clients with egg-freezing providers across the country. I arrived a few minutes early at the designated room in a midtown office building and found several dozen women with expensive handbags and pinched foreheads quietly clutching their Chardonnays and thumbing through slick folders. I didn't want to talk to anyone either and was grateful no one approached me. What would we say to each other? "So you're on deadline too?" Part of me wanted to know, though. How *did* we end up here? We all looked as if we belonged at a charity luncheon, not a seminar on how to rescue our last bits of fertility.

The first speaker was Dr. Alan Copperman, a reproductive endocrinologist from the fertility clinic Reproductive Medicine Associates, which was the New York City partner of Extend Fertility. He showed us the familiar depressing graph illustrating how chromosomal abnormalities in women's eggs rise with age and increase the chances of birth defects and miscarriage. As I watched the audience's brows furrow even deeper, I knew we were going to need a lot more Chardonnay to make it through the night. One woman looked as if she was about to burst into tears.

Then Dr. Copperman pointed to a flat line representing eggs donated by young women that were used to make embryos and transferred to infertile patients; the rate of chromosomal abnormalities remained the same, even as the age of the recipients rose. That's why the popularity of donor eggs had grown over the past two decades; doctors could offer such young eggs to patients who no longer had good eggs of their own. By freezing our own eggs in time when they are still viable, he explained, we could in effect "donate" them to ourselves years later.

Copperman acknowledged that the early success rates of egg freezing were disappointing, but he said the science had since vastly improved. Although the procedure was still considered experimental (translation: not an established medical practice and requiring informed consent from patients), as of September 2006 some four hundred babies had already been born from frozen eggs around the world. In fact his clinic had recently conducted a small study that showed three out of four patients had become pregnant from frozen donor eggs. "This technology is ready for prime time," he said. "We can make the curve straight."

Hands shot up with questions about the success rates and cost. As for the chances of the procedure working, Copperman explained that it depended on when women froze their eggs: you had a "great" chance of success if you froze before age thirty-five, whereas you could expect a "reasonable" chance of success if you froze before forty. He didn't recommend the procedure for women any older. The price was about $10,000 to $12,000, plus several thousand more for the drugs, and several hundred more annually for storage. Part of the fee went to Extend Fertility, which coordinated storage. If your body produced a low number of eggs, you might need to undergo another round to make sure you put away enough.

A lot of the women looked as confused as I was. How were we supposed to interpret "reasonable"? And was "reasonable" worth that much money? Copperman reviewed the IVF success rates with fresh eggs at his clinic: a healthy thirty-five-year-old had a more than 50 percent chance of IVF working; by age forty, that rate dropped to 30 percent. You had to assume that success rates for frozen eggs would be lower, since some eggs wouldn't survive the freezing process or fertilize at all.

I still didn't know what to make of these odds. Then the next speaker, Dr. Georgia Witkin, the clinic's house psychologist, told us what the chunk of cash really bought us: a sense of control over our lives. "It's just in case," she explained. "It's not that women plan to use them. They just want to know they've done everything they could."

A rep from Extend Fertility added that several of their two hundred clients had married and become pregnant on their own after freezing their eggs. Perhaps both women *and* men relaxed once they were no longer under the gun to make a decision. Was freezing a mechanism that moved relationships forward? Did it work in the same way that telling a man you weren't looking for a committed relationship supposedly inspired him to want one, according to several relationship advice books?

By the time I left the seminar, I was no longer trying to analyze success rates. I was thinking about the implications this would have for my relationship with Paul. Witkin had acknowledged what we all knew well: the biological clock puts unwelcome stress on a relationship. I wondered if Paul and I would be having a different conversation about children if we had met five years earlier, when my fertility wasn't front and center. Surely, having a family would seem more natural then. Perhaps egg freezing could make up for our meeting each other so late in life.

I waited until the weekend to tell him. I was pretty sure he would support me in whatever I pursued, but I was still nervous. Would he interpret the conversation as removing the pressure or forcing the issue? After several sips of pinot grigio during dinner, I eased into the subject by telling him about the seminar as part of my reporting assignment.

"Sounds interesting," he said.

"I was thinking it might make sense for me to do it too," I said.

It turned out Paul *loved* the idea of egg freezing. "It does make sense," he responded, looking visibly relieved. "Options are a good thing," he added. "I can't tell you how many dates I've been on when, before we've even finished the salad, the woman says, 'You know I've spent so much time on my career that now I want to focus on my personal life. I now know what I want.' What are guys supposed to do with that?"

He had a point. Then he had an even better one. "How many people do we know who are in unhappy marriages because they thought they were running out of time? Where do you think some of them would be today if the women had known about egg freezing?" I knew of three couples in their mid- to late thirties who married because they were pregnant. Presumably they were open to the idea of children or else they would have been more careful with birth control, but the women had admitted to me that fertility pressure clouded their judgment. They also said they wished they would have taken the time to get to know their partner better.

"I think you and I are doing it right by going slowly," he said confidently.

"I think we are too," I replied.

When I told my mother about my freezing plans during a trip home to San Diego the following week, she remained quiet for several uncomfortable minutes as she washed the dishes. "Wouldn't it just be easier to get married and pregnant?" she finally asked. "Why are you making this so hard?"

"I'm not doing this for Paul," I insisted, trying to head off the next question. "I'm not ready to have a baby, either."

"No one is ever ready," she replied. "You'll do just fine. You'll be an excellent mother. Now, why don't you date someone who wants to be a father?"

I quickly changed the subject. "Even if I got pregnant tomorrow, I still might not have time to have a second baby," I pointed out. "Wouldn't it be useful to have some eggs put aside?" I also explained that it was a good investment, as I wouldn't have to spend tens of thousands of dollars on IVF cycles later.

"I suppose so," she conceded.

I had expected some pushback from my mother, who was anxious to see me settled down. However, I was surprised when my friend Heather, who was single and despondent about turning thirty-nine in a month, turned up her nose at the idea. After all, I had fielded many phone calls from her over the years about how her dates seemed to lose interest once she told them she wanted marriage and a family.

"Can you imagine what it would be like to date without this fertility thing hanging over you?" I asked. "You know guys have to sense it."

She said she couldn't afford the procedure. Besides, she was scared of giving herself shots.

"You can get a loan," I offered. "And I can give you the shots."

"I don't know. I'm just so nervous," she responded. "The whole thing is freaking me out."

Truthfully I wasn't completely sold on egg freezing myself. Some days the idea brought a wave of relief. But another part of me worried that I wanted it so badly that I had lost the ability to tell if I was making a good decision. Then there was the cost, which would pretty much wipe out my savings.

But I was tired of being freaked out. An issue that had once seemed amorphous and scary now seemed clear and manageable. All I had to do was follow instructions: choose clinic, pay bill, make appointments, take drugs, and show up for retrieval. Egg freezing gave me something to do.

Although the medical community designates women over thirty-five as "advanced maternal age" and recommends extra monitoring and birth defect screenings, many experts told me that the decline of women's fertility really accelerates after thirty-eight. Considering I would be thirty-six and a half soon, that meant I had about two years left to put away some eggs before they went to pot. I definitely didn't anticipate having a baby before then. If things went nowhere with Paul, I would still have my eggs, and I could decide when and with whom I would use them. If my eggs didn't work, I still had my list of ways to be a mother taped to my desk lamp.

Freezing my eggs seemed like the best decision I could make with the information I had at the time. Just as Dr. Witkin had said, I would know I had done everything I could.

Christy Jones

In the spring of 2002, Christy Jones was living back in her parents' home in Santa Barbara, California, trying to figure out what next to do with her life. She had already graduated from Stanford, cofounded a software company, earned 2 million frequent-flyer miles, and made the cover of *Forbes* magazine three times.

One evening at dinner her mother announced out of the blue, "All the girls in the Jones family should freeze their eggs when they turn thirty." She flinched at her mother's age cutoff, since it was clear she was referring only to her three younger sisters. Jones was thirty-two and didn't have much hope of producing a grandchild soon, especially since she and her boyfriend of four years had decided to take a break.

Her mother described her chat earlier that day with a friend who had just seen the author Sylvia Ann Hewlett interviewed on *Oprah*. Jones didn't need any elaboration. She knew all about the author's message that career women were woefully ignorant about when their fertility ended. The subject was in the news and on the lips of many of her anxious girlfriends. It had sent Jones into a tailspin about her own biological clock, since she was applying to business school and would be thirty-five by the time she graduated.

Jones had always assumed she would be married by her early thirties. Even though her mother had frequently encouraged her to find a husband by her mid-twenties if she wanted a family, she hadn't become serious about dating until she was twenty-nine, when she met her boyfriend Rob through a mutual friend while they were both working in

Austin, Texas. She never had the time to date: after graduating from Stanford with a degree in economics, she couldn't pass up opportunities that were unheard of before the Internet boom and had spent most of her twenties working grueling hours to grow her business. What started as a summer job after her sophomore year had led to the creation of the hugely successful software company Trilogy. At age twenty-five, she spun off a division called pcOrder that sold software to help computer manufacturers sell their products online. Five years later, she had taken pcOrder public and sold it back to Trilogy for more than $100 million.

Like her mother, Jones had been curious about using egg freezing as a way to safeguard her own fertility and was disappointed when she read in Hewlett's book that human eggs were too fragile to survive freezing and thawing. Why were scientists able to freeze sperm and embryos but not eggs? Jones wondered if there was a business opportunity in egg freezing.

When she did an Internet search about egg freezing in her childhood bedroom in 2002, there were thirty published births, though some sources claimed the unofficial number was closer to a hundred. According to an article in the *Los Angeles Times*, six egg-freezing programs already existed in the United States, and many more doctors planned to offer the procedure soon.

Jones had never forgotten a statistic Hewlett had included in her book: the majority of the "high-achieving" women over forty-one without children didn't choose their status. In fact more than a quarter of that group said they would still like to have children. She couldn't ignore the fact that as she and her friends approached their mid-thirties, it was likely that many of them would be included in that statistic.

Jones knew from years of start-ups that there was a magic moment to launch a company. Even though egg freezing was considered bleeding edge, there was already an incredible need for it. The women over forty described in Hewlett's book had been caught off guard because many were ignorant about the limits of their fertility. But Jones's peers couldn't ignore the stark reminders that filled women's magazines and talk shows. The difference was that *these* women still had a few years to put away some eggs. She reasoned that fertility doctors would have more clinical experience by the time women actually wanted to thaw their eggs several years down the road. The more she learned about the work of researchers in Atlanta and Jacksonville, the more she realized the trend had already begun.

She came across a link for the Stanford IVF Center advertising that egg cryopreservation was available to cancer patients who wanted to protect their eggs from damage during chemotherapy and radiation treatments. The following week, she was in the office of Barry Behr, the clinic's embryologist. "There's such a big market of women who would be interested in this," she said. "Is there a future in egg freezing?"

Behr was skeptical, but he knew of a few scientists who had done some small experiments in their labs and had achieved a pregnancy rate of nearly 30 percent, which was similar to the rate of many clinics using frozen embryos. He also told her about an Italian scientist in Bologna who had developed a chemical solution a few years earlier that improved egg survival rates dramatically, to more than 80 percent. Jones looked up the study and learned that nearly 60 percent of those eggs had been successfully fertilized. Although those numbers were encouraging, she quickly realized that not all of those embryos would continue to develop, and only a handful would be implanted. Egg freezing still had a long way to go.

In the meantime, Behr conducted his own research and found that the eggs that had been frozen and thawed the Bologna way developed into embryos almost as robust as those made from fresh eggs.

Later that spring Jones was accepted at Harvard Business School. She planned to use some of her time during the program to think about how to bring egg freezing to market. She even flirted with the idea of opening up her own clinic. To make egg freezing seem less scary, she wanted the clinic to feel more like a spa than a medical facility and envisioned a sleek design with white walls, lots of light, big vases of orchids, and turquoise pillows. She imagined packaging the service with a chic hotel or limo service as part of a girls' weekend, so women could support each other during the procedure.

But she quickly abandoned the idea when she looked at the numbers. With doctors' and nurses' salaries, equipment, and infrastructure, she would need at least a hundred patients a year to break even. Although she expected high demand for egg freezing, that seemed like a lofty goal for a new practice. As she thought about how to develop a new business model, she remembered a company that sold designer braces. She had always wanted to fix her slightly crooked front teeth but didn't want to wear big metal braces when she spoke in public. When she saw ads for Invisalign, which makes removable clear plastic aligners, she was intrigued enough to go to its website. There she found tutorials, patient testimonials, and a listing for a local orthodontist who offered the product. She liked the way the company educated consumers directly about a new product rather than relying on dentists to tell their patients.

Fertility clinics needed such a bridge to their clients, she thought. Potential egg-freezing patients weren't seeking medical help to become pregnant and had no reason to visit a fertility clinic, so they wouldn't necessarily learn about egg freezing from fertility doctors. Such a niche group required specialized marketing efforts.

That summer Jones did some research with her classmates, figured out the specifics of her business model, and set out to win the Harvard business plan competition. Her company would handle the logistical side of egg freezing: soliciting and educating clients, transporting and storing eggs, supplying the freezing solution, and training embryologists. She would build a national network and call it Extend Fertility. Patients would pay one fee; the lion's share would go to the fertility clinics performing the procedure, and Extend would take a cut.

Colleagues warned Jones that doctors were notorious for wanting to control their own empires and might not be receptive to her overt commercial approach. But she believed doctors could use help building awareness about the need for and availability of egg freezing. She also argued that she was bringing them the science. She would arrange for Raffaella Fabbri, the Italian embryologist responsible for the breakthrough in the freezing solution that led to higher egg survival rates, to travel to their clinics and show them the most recent protocols. Extend would also supply the freezing solution, which Fabbri had patented with a Danish company. As for research results, Jones offered to do studies jointly with doctors. She had approached pharmaceutical companies to secure medical grants to pay for stimulation drugs. How could doctors not be intrigued?

Like her potential clients, Jones didn't have time to wait for the perfect moment when doctors had "enough" clinical experience. It made sense to put some eggs in the freezer now. She was ecstatic when Behr retrieved twelve of her own beautiful eggs and whisked them to the freezer.

In the spring of 2004 Extend Fertility became a real company. Although Jones had won Harvard's business plan contest, she was unable to convince venture capitalists to invest in such a small enterprise. So she used her own money from the sale of her software company and went to work creating a website, hiring a representative to answer inquiries, convincing doctors to join her network, and assembling a scientific advisory board. Behr, who ran the Stanford clinic and was also the scientific director at Huntington Reproductive Center in southern California, announced that both these clinics would become Extend's first partners.

Next Jones had to find patients. During a hike with some college friends, she was introduced to a freelance journalist who wrote for *Elle* magazine and wanted to do a piece on Extend Fertility. Jones had never been featured in a women's magazine before, but at age thirty-four she made a sympathetic spokeswoman for her demographic. Although she initially came across as polite and polished, especially when she tied her lustrous auburn hair into a neat ponytail, the furrow in her brow made her seem vulnerable. Women knew that Jones worried too.

The media jumped on the story. When the *Elle* article hit the newsstands in April, a producer from *Good Morning America* invited her on the show. By the end of the week, two dozen women had called Extend wanting more information. Jones and a nurse she had hired part-time answered calls from a Harvard conference room. They received a hundred inquiries in the first two months and signed up thirty women for the procedure.

At first, Jones was thrilled with the exposure, which lauded egg freezing as an important advance for women. In fact the media—and many clients—seemed to overlook the fact that Extend Fertility and its affiliate clinics hadn't actually frozen any eggs of clients other than Jones and a handful of cancer patients. Nor had they thawed any eggs to make a baby. But her critics took note. In one article in the *Chicago Tribune*, Dr. Kevin Winslow, who froze eggs in Jacksonville, Florida, was quoted as saying that Jones's idea of building a network of clinics whose doctors had little experience freezing eggs was "absurd and irresponsible." Jones had expected opposition, which she knew was the price of publicity, but the tone of the attacks took her by surprise. In Atlanta, Dr. Michael Tucker told reporters that it was unethical to offer egg freezing to patients when there wasn't enough research data. "It is truly just taking people's money on a wing and a prayer, hoping things will work out down the road," he was quoted as saying in the *Austin American-Statesman*. Lori Andrews, a lawyer specializing in biotechnology, told the *Chicago Tribune* that egg freezing played off women's fears that motherhood was passing them by: "Really what we need is not a Harvard Business School student coming up with a business plan to make money by holding out hopes that have not been scientifically proven."

Condemnation of such commercial interests was also a dominant theme at the annual meeting of the American Society for Reproductive Medicine in 2004 in Philadelphia. Some of egg freezing's fiercest critics called the procedure risky and a long shot at best. They pointed out that no one knew the long-term health risks to babies born from thawed eggs, especially given concerns that freezing could damage an egg's chromosomes. And what would happen to all these women who were counting on their frozen eggs to work and would be bitterly disappointed when the eggs failed and they were too old to get pregnant on their own? The professional society, which had no authority over doctors but was influential nonetheless, even released a statement saying that egg freezing was experimental and shouldn't be offered or marketed as a means to "defer reproductive aging."

The outcry even drowned out promising news about egg freezing. At the same meeting, Italian gynecologist Eleonora Porcu, who along with Fabbri had achieved the first birth in the world from frozen eggs using ICSI seven years earlier in 1997, reported thirty-three more births of children conceived the same way and brought the world total to more than 300.

Jones was aware that she needed research results in the form of babies to gain credibility. After years of talks, she managed to convince the pharmaceutical company Serono, which made egg-stimulation hormones, to earmark $500,000 for clinics that were interested in conducting studies. She added some of her own funding and approached doctors at Reproductive Medicine Associates in New York City about doing a joint project. (It was also part of her strategy to sign up a respected clinic in the city, a mecca of professional single women.)

In May 2005 Extend and Reproductive Medicine Associates launched their first study. Out of seventy-nine total eggs, sixty-eight survived freezing and thawing and sixty-one were fertilized. Approximately one-third of the embryos, about six per patient, were transferred, and three of the four women become pregnant, two with twins. They had five babies from nearly eighty eggs.

3. *Reprieve*

Monica

Monica's coworker Gary surely had no idea of the chain of events he would set in motion the day he asked this simple question: "What did you do this weekend?"

"I went on another date," Monica began in her usual Monday morning report, which quickly turned into a rant. "I'm sick of this whole thing being so much work. These guys are great on paper, but there's zero chemistry when you meet them. Then you have to start all over again."

Gary tried to commiserate and commented that his wife's friend was also in her late thirties and having a hard time finding a husband. "She's so fed up that she told my wife she's thinking of freezing her eggs," he said.

Monica had read about egg freezing more than a year earlier in 2004, but this was the first time she heard of someone actually considering it. At thirty-five, she hadn't thought about it for herself because she assumed she'd be married soon or at least well on her way. She never dreamed she'd still be single at almost thirty-seven. Despite her hectic schedule, she managed to set up about one date a week, but it never went anywhere. One guy appeared annoyed when she told him she would be out of town for the following three weeks on business trips. Some men living in New York City complained that she lived about an hour away in central New Jersey. One guy took her out six Fridays in a row but never tried to kiss her. Monica declined the seventh date; she wanted a boyfriend, not a dinner companion.

On long flights home, she wondered if she was putting enough time and effort into finding a husband. Maybe the Friday guy would have worked out if she hadn't been so impatient. She knew dating was a numbers game. You only had to connect with one person. She just didn't know how long it would take to find *him*.

Unable to ignore her curiosity, she researched egg freezing on the Internet that afternoon at work, and a few entries popped up on Google, mostly media articles with scary statistics on age and fertility, a couple of doctor advertisements, and several cooking sites offering instructions on how to freeze chicken eggs. Near the top of the list was an entry for a company called Extend Fertility. She clicked on the link.

Monica had a hard time accepting that her eggs were deteriorating inside of her. She didn't know whether to credit luck or her Filipino genes, but at thirty-six, her body showed few signs of aging. Her skin was still smooth, and her dark eyes and hair were shiny. Her petite body looked fit and toned, even though she rarely broke a sweat or watched what she ate.

She was enthralled by the idea of stopping her eggs from aging. Compared with the messy business of falling in love and synchronizing timing and expectations, egg freezing seemed wonderfully rational and proactive. Surely she would be married within a few years and have enough time to get pregnant. But then she thought she'd be married by now when she canceled her engagement three years earlier. Securing a backup appealed to her very prudent core.

But first she wanted to know what her parents thought of it. Since both were doctors, he an ER doctor and she a psychiatrist, Monica trusted their medical opinions. She also wanted to see if they were willing to help her pay for it. Her parents were comfortable financially, and they had a history of helping their kids with big investments, such as school tuitions, cars, and house down payments. She called her parents that afternoon and asked them to meet her the following week for lunch.

Monica intended to check out Extend Fertility's partner, IVF New Jersey, but she also booked a consultation with a reproductive endocrinologist at Saint Barnabas, a large health care system in New Jersey. The doctor there explained that he had frozen the eggs of several cancer patients but did not offer the service to healthy women. Monica left his office frustrated. Why wasn't she too allowed to protect her ability to have babies?

She met her parents at a Chinese restaurant near their home in Princeton. "It's very important for me to have kids," she began. "And I'm not really dating anyone. I've heard about a technology that lets you freeze your eggs so you can use them when you're older. I want to know if you think it's valid."

After a long pause, she added, "I'll need help with the cost, which is around $12,000. Of course, we could view it as a loan."

Without missing a beat, her dad took out his checkbook while her mother launched a fusillade of questions about the success rates. Apprehensive that her mother would poke holes in her plan, Monica felt her shoulders tighten. She knew it was wise to be cautious about any technology that claimed to be revolutionary, but she wasn't willing to trash the whole idea simply because it was new.

Still, she took a deep breath and told her parents the facts that she was sure would sink her case: that just a handful of babies had been born from the procedure and that the clinic she was considering was part of a network that had frozen the eggs of only a couple dozen clients. Then she braced herself to hear her mother say that the science wasn't established or that surgery was risky, but instead she was encouraged when her mother wanted to know more: What happened to the eggs if she didn't use them? How did they discard them? Her father tossed a blank signed check across the table, and Monica invited her parents to her consultation with Dr. Susan Treiser of IVF New Jersey.

Dr. Treiser warned Monica that after the thawing and fertilization of her eggs, she would end up with only a handful of embryos to be transferred to her uterus. However, if Monica produced a lot of eggs, she could have multiple chances. Because data were so scarce, there was no way of knowing her odds of success.

Monica figured that the technology would improve by the time she thawed her eggs and scheduled her blood tests and ultrasounds. A few weeks before she turned thirty-seven, Dr. Treiser harvested nineteen eggs. "That's it? It's over?" a groggy Monica marveled as she curled up against the backseat of her parents' car. She felt a certain pride that she had produced more eggs than most women her age and well above the fifteen the doctor had recommended. As she looked out the car window, she savored a sense of security as her parents took her back to their house to stay in the bedroom she had used as a child, back when she believed everything was going to be okay.

Her chest pains returned a few months later. Since her episode in Minneapolis the year before, she had seen her family doctor and ob-gyn, but their tests had been inconclusive. Sometimes the pains showed up several months in a row, and sometimes they stayed away for months at a stretch. But Monica had noticed that they usually occurred around the time of her period. At her family's shore house later that summer, she mentioned the coincidence to her mom.

"Monica, do you know what a period is?" her mother asked in the psychiatrist voice that had made Monica bristle ever since she was a child. Before Monica could respond, her mother provided the answer. "It's an unrealized pregnancy."

Her mother suggested that the chest pains were a manifestation of Monica's subconscious anxiety about getting older. "Maybe it's the way your body is choosing to react," she offered. In other words, her body was frustrated because she had yet to have children. Since the doctors couldn't find any physical cause, the pains were likely psychological. Monica quickly dismissed the theory. If her mother was right, why didn't her body know she had bought it some time? Why was it still hell-bent on reminding her that her fertility was fading, like some last-minute warning system? Egg freezing had brought short relief.

Kelly

Dr. Kevin Winslow sat in his corner office with wide views of downtown Jacksonville. He was wearing a crisp white lab coat and waving around a recent issue of *People* magazine. "Having a Baby after 40," read the cover, which featured a radiant Geena Davis.

"See these women?" he lectured Kelly, who, after much nudging from her friend Courtney, finally had scheduled a consultation. "The vast majority didn't get pregnant with their own eggs. They had to use eggs from a donor, but the magazines never mention that." Kelly already knew that having a biological child in her forties could be challenging, but she had presumed success was simply a matter of having enough money and seeing the right doctors. How could Hollywood stars have trouble?

The words *donor eggs* rang hard in her ears, until she reminded herself that she had flown to Jacksonville to learn about freezing her eggs to avoid having to use them. She had just turned thirty-eight a few weeks earlier and was proud she had made it into Winslow's office almost a year before his cutoff age. Still, even though she was technically eligible, Winslow called her "borderline" and said she needed a blood test to make sure she hadn't gone into early menopause.

Winslow explained that she could expect to produce eight to ten eggs at her age. (A forty-year-old might produce just five eggs, and the quality might not be as high, he said.) Kelly likely would need to undergo a second round to produce the desired twelve to thirteen eggs that he advised women thirty-six and older to put away. That would give her enough embryos for two tries. For example, he might thaw six eggs at a time and end up with two embryos to place into her uterus. According to his clinic's latest data, she had a 30 to 40 percent chance of each transfer (of several embryos) resulting in a pregnancy.

Kelly left Winslow's office carrying a large folder and feeling over-whelmed. How would she find a spare $10,000? What did it feel like to take so many hormones? What if she took all the shots and didn't produce enough eggs? What if the eggs she did produce weren't of good quality?

She also felt profoundly grateful that her coworkers had cared enough to encourage her to look into egg freezing. She thought that other women needed to be educated about the stark facts of their fertility while they still had time to preserve it. Maybe she would ask Winslow to come to Charlotte and give a lecture to her professional women's group.

Kelly didn't know at the time that for many women, egg freezing wasn't actually an option. Although Winslow acknowledged that the procedure was experimental, he talked about it in such a matter-of-fact way that she assumed it was a standard service at all fertility clinics. But in 2003 Winslow was one of only a handful of specialists in the world offering egg freezing to women who didn't have a compelling medical reason to use it, such as when a patient loses her fertility during cancer treatment.

Winslow and his Chinese-born embryologist, Dunsong Yang, made their small Jacksonville clinic a leader in babies conceived with frozen eggs in the United States. With twenty-five babies (including three sets of twins and one set of triplets), they were responsible for a quarter of the world's hundred or so documented cases.

Winslow had always been interested in egg freezing, which he called the holy grail of fertility medicine, but it was Yang who started the program in Jacksonville. In 1996 Winslow needed an embryologist to increase the size of his practice and lured Yang from rainy London to Florida. (The higher American salaries and big homes with swimming pools didn't hurt.) Five years earlier, Yang had left China for Britain after becoming frustrated with his country's lack of interest in reproductive science and limit of one child per family.

In London Yang pored over medical journals at the British Library and studied Debra Gook's experiments showing that human eggs could survive freezing and thawing. He thought her results could be improved by using a higher concentration of sucrose in the solution to expel more water from the eggs before freezing them. Also, if he prepped the eggs for freezing at a higher temperature, which was closer to one's body rather than room temperature in which embryos were routinely frozen, the solution could enter the egg quicker and better protect the egg.

When Yang announced to Winslow and his colleague at the time, Dr. Patrick Blohm, that he wanted to try his hand at freezing eggs, both doctors seemed skeptical. No embryologist had been able to duplicate the first reported egg-freezing pregnancies from a decade earlier. But Yang received approval for his research from his hospital's review board, and Winslow gave him six eggs left over from a twenty-nine-year-old donor.

Yang froze three eggs with his own formula and three eggs with the conventional method used to freeze embryos and waited for a patient. Several months later, a nurse approached Yang and reported that a forty-three-year-old patient wanted to use donor eggs but was having trouble finding fresh ones. "Do you want to try your frozen eggs?" she asked him. The patient signed a consent form, and Yang thawed the eggs. The ones frozen the conventional way were duds. But two of the three he had frozen using his new recipe survived and successfully fertilized.

Two weeks after Winslow transferred the embryos to the patient's uterus, he called Yang with the results of the pregnancy test. "It's positive!" he announced. Two weeks after that, they learned the patient was having twins. By 2002 Yang's team had produced sixteen frozen-egg babies. As long as they obtained hospital approval, they could offer egg freezing to patients like Kelly on an experimental basis.

For the next few months after her consultation, Kelly still couldn't believe she was being forced to make a decision so quickly about preserving something she never thought she was losing in the first place. In the meantime, she didn't want to push her boyfriend, Steve, to overcome his reluctance to have more children, and she couldn't work up the nerve to break up with him. But the white elephant in the room made it hard to dream about a future together, and the relationship had started to feel stagnant. They talked about marriage only casually and avoided the topic of kids, which usually led to awkward conversations that never made her feel any better. The topic of egg freezing was just as uncomfortable.

Finally, one evening, Kelly forced herself to bring it up. "You know that I want kids, and I've heard about an option that might make this possible down the road," she said. She explained the process, and Steve responded, "That's a great idea. You should look into that."

As the months wore on, Kelly became increasingly aware she was paying a price to hang on to Steve. She enjoyed living in the moment, but she was also squandering the opportunity to meet someone she *could* marry and have children with. So, in a moment of courage and clarity, she invited him over for homemade lasagna and delivered the news. "This probably isn't going to come as a surprise, but I don't think we should see each other any more," she said.

Steve was silent. As tears welled in his eyes, he gathered the few belongings he kept at Kelly's house, kissed her, and left.

Kelly sobbed for the next two days. She had become attached to his daughter and felt awful that she would no longer be a part of her life. Though she had left Steve to free herself to meet someone else, in the following months she didn't feel much like dating. Rather she focused on house projects and rekindled her love for gardening and working in the yard. She also reconnected with friends she had seldom seen while she was with Steve and concentrated on expanding her social life.

She didn't feel pressed to return to Dr. Winslow's office. Although she knew that her fertility was dwindling, she felt as if her deadline of age thirty-nine was several safe months away. The only pressure she felt was from her colleague Courtney. "You waiting on something to happen here?" she'd ask Kelly during lunch. "If you're gonna do this, then just do it!"

"I'm still making a decision," Kelly replied. "Just give me a couple more months."

Why *was* she waiting? She told herself she still needed time to figure out how to pay for the procedure. Since her divorce three years earlier, she had focused on creating the financial stability that had been missing in her marriage. She had found a job with a decent salary, paid off her debt, bought a house, and even stashed away about $14,000 in the bank. She was proud of her savings, and the thought of seeing the bulk of it disappear in one big check seemed unthinkable.

But she knew she could find the money. The real reason she hadn't scheduled her blood tests to measure her hormone levels was because she was terrified of the results. What if she was already infertile? She wanted to put off bad news as long as possible.

Kelly saw Dr. Winslow again later that fall, when he gave a talk to her professional women's group. About two dozen people attended, and Kelly was proud to hear such an open discussion on a private topic. After the presentation, she approached Winslow. "I know I have to get moving on egg freezing," she admitted. "I'm going to be thirty-nine in a few months. Will you still see me?"

"I'll still see you," he replied. "But you need to call for an appointment ASAP."

"I will," she said confidently. "I will."

On her birthday, January 13, 2004, Kelly's alarm finally sounded. "I'm thirty-nine, not married, and not even dating," she cried to Courtney one morning shortly after. "I've got to do something!" The slow-motion denial she had embraced during the past year was now replaced by panicky urgency. Now that she had officially missed Winslow's age cutoff, the fact that she might have lost her chance to have a baby suddenly seemed unbearable, especially when she thought about why: Waiting out a bad marriage. Hanging onto an ambivalent boyfriend. Avoiding test results. They were all terrible reasons. Kelly was ashamed that she had been such a poor advocate for herself. She wanted those eggs out of her as soon as possible.

She was so nervous when she called Dr. Winslow's office that her heart raced. "You might have blown it," she admonished herself. "He might not even see you!" But she hadn't blown it, and her eyes welled with tears of relief when he said he would make an exception since her birthday was so recent. She thanked him profusely and set out to find the money.

Kelly couldn't bring herself to plunder her savings. She considered taking money from her retirement account, but she didn't want to face the stiff penalties. Then she realized the answer was sitting in her junk mail. She usually received four to five credit card solicitation letters a week, some promising no interest for a year. She often fantasized about applying for one and buying new furniture for her house. She figured she could use one to finance her egg freezing and pay off the balance throughout the year. "This is one expensive lottery ticket," she thought as she filled out the application. Her family told her she had a made a good decision. It was a welcome change in conversation from her jokey threats to visit a sperm bank if she never remarried.

More than a year after her initial consultation, Kelly finally had her hormone levels tested and was ecstatic to learn that her follicle-stimulating hormone (FSH) level was 8.9; Winslow considered any number less than 10 to be desirable. A high number meant she didn't have many eggs left and that her pituitary gland had to work overtime to signal her ovaries to pop out eggs. (The ovaries of a fertile woman react more quickly.) It also meant she wouldn't respond well to stimulation drugs. Doctors also check a patient's level of estradiol, which is a form of estrogen that indicates how well her ovaries work and validates the FSH result.

Although Kelly was told she might need to undergo two cycles to put away enough eggs, she knew she couldn't afford a second and resolved to do this round perfectly, as if being a good follower of directions could make up for years of not paying attention to fertility advice. When a nurse showed her how to mix and administer the medications, Kelly worried she wouldn't take the right dosage to get enough of the drug in her body to make her eggs grow. She took home a folder with several handouts and charts and reviewed the instructions several times, often paging the nurses with questions: Was she reading the measurements on the syringe correctly? Was it okay if she took one injection an hour late? Did it matter if she used a different size needle than the one they recommended? Was it normal to be so bloated?

Near the end of her cycle, Kelly was so fed up with giving herself shots that she rejoiced as she finished up her medication just in time for the scheduled retrieval. Her older sister drove nearly six hours from Atlanta to be with her for her last ultrasound. Kelly lay on the examining table feeling triumphant as the doctors counted and measured all her follicles, at least a dozen in all. But a few hours later, her spirits sank when Dr. Winslow asked her to continue taking her medication for one more day. He wanted to give her smaller follicles a chance to catch up to the larger ones so they could harvest as many eggs as possible. Another day meant about $600 in additional drugs. Kelly reeled when she calculated that the drugs alone cost around $4,000 and brought her total bill to nearly $14,000.

Finally retricval day arrived. At age thirty-nine and seven weeks, Kelly hoisted her legs on the operating table in Winslow's office and surrendered to the anesthesia. When she came to, his staff was standing around her clapping. "We got nineteen eggs," one shouted. "That's incredible!" That number was double what Winslow had expected and well above the dozen he had suggested she freeze.

"I *knew* I was still fertile," she wanted to say to them. "I don't know what all the fuss is about."

Hannah

For a few dark months after the breakup with Marcus, her boyfriend of six months, Hannah was consumed with despair. At age thirty-eight, she worried that she wouldn't meet someone else in time to marry and have a family. However, the end of her longest relationship ever confirmed an even worse fear that had nothing to do with her biological clock: Hannah was convinced she was unlovable. Unlike other women who occasionally wrestle with self-doubt, she believed she had a personality defect that made it difficult for men to connect with her. She knew she was pretty and intelligent, but she had a hard time feeling comfortable around men. As a result, they desired her, but they didn't fall in love with her.

Every fizzled relationship and breakup stung worse than the last and, in Hannah's mind, provided evidence of what was becoming an incontrovertible truth. She didn't bother summoning the usual post-breakup rationales. She didn't try to blame bad timing, his commitment fears, or a lurking ex-girlfriend. The breakup with Marcus *was* about her.

Hannah had been seeing a therapist for the past couple of years, but she felt as if the sessions consisted of her complaining and the therapist nodding her head: comforting, but not helpful. In the meantime, she received a referral for a new therapist who used a different approach, called cognitive-behavioral therapy. Rather than console her, he challenged her to question her negative thinking. He asked her to keep a list of thoughts that made her feel bad; she then had to consider each one and ask herself if it was true. For example, if she thought "I'm unlovable," the therapist asked her to tally up all her friends and family who did love her. If she thought "I'll never get married," she had to remind herself that she was in treatment to learn skills to prepare herself to meet someone, or she had to remember that plenty of nice guys had been interested in her over the years but, as her mother often said, she didn't like those guys.

The technique seemed simplistic, but it was surprisingly effective. It wasn't so easy to throw around seemingly foregone conclusions, such as "I'm a loser," when you couldn't find evidence to back them up. In the beginning, Hannah could escape her hateful inner voice for minutes at a time before giving in to familiar cruel sentiments, such as "I'm thirty-eight. What makes me think I can pull off getting married now? I'm going to be alone for the rest of my life." Over time, the relief lasted longer, sometimes even days, and she could come back from a hopeless place more quickly. She also attended group therapy, learned relaxation and meditation techniques, and gradually started to feel better.

Hannah knew she had a lot of work left to do before she was ready to be in a long-term relationship, so in the meantime she thought she should focus on creating a satisfying single life. She felt a certain pride that she hadn't put some of her dreams on hold, such as buying a home or traveling, while she waited on love. And she'd always managed to fill up her time planning activities with friends or family.

But as she approached her late thirties, she found it harder to be single in what had become a couples world, since many of her friends were married or in serious relationships. At first, she didn't mind assuming the role of third wheel, but she noticed that when her friends had kids, they started going on trips with other families. She appreciated the way they tried to include her in their lives by inviting her over for dinner and promising "You'll always be Auntie Hannah." But she often left their homes feeling even more alone. There was so much she was missing out on.

Hannah liked to think she wasn't attached to the idea of having a baby. Her ambivalence was partly protective; if she didn't let herself get in touch with her desire to be a mother, she couldn't be disappointed if she never became one. Having low expectations was a coping mechanism she had carefully refined to handle years of romantic heartache. It was easier to try to be happy with being single and forget about weddings and children. It was easier to say you were the kind of person who simply wasn't *into* those things.

Another reason for her uncertainty about motherhood: she was scared she wouldn't be good at it. Although she was usually warm and caring, she froze up around children. She admired the way Lori, her best friend from college, could effortlessly scoop up infants and talk to toddlers. Hannah, on the other hand, had held only a handful of babies in her lifetime. She didn't know what to do if they started to cry and quickly handed them back to their parents. Her discomfort further fueled her fears about being unlovable. What if a baby didn't love her? Or worse, what if she couldn't love a baby?

Only when she became an aunt did she let herself imagine becoming a mother. She watched carefully as her sister-in-law took care of her little nephew, Tyler, and learned basic tasks, such as how to change his diaper and soothe him when he was distraught. When Tyler cried, Hannah picked him up, patted his back, and said, "There, there. It's okay." She was moved by how powerful she felt consoling a baby. She enjoyed reading to him his favorite stories about Thomas the Tank Engine and Barney the Friendly Dinosaur. She loved watching her brother egg him on to say hard words, such as *millionaire*, and laughed when he tumbled over the consonants. When her family went out for dinner, she would carry the baby around the restaurant and show him the aquarium and quiz him on the colors of the fish. But her heart softened most when he crawled into her lap to cuddle. Tyler loved her, and she loved Tyler.

During such moments, Hannah reluctantly admitted to herself that she really did want a family. She had formed such a strong relationship with her mother that she longed to develop the same bond with her own child. On occasion, she even indulged her dream. While looking for baby gifts for friends at the Gap, she bought orange cargo pants for a future baby. In France she shopped for some pretty lime-green yarn and a knitting pattern for a tiny cardigan. On that same trip, she found matching felt shoes. She stashed them all in her hope chest.

Hannah grew more concerned about her biological clock. After reading that wildly fluctuating menstrual cycles were a sign that a woman's fertility was declining, she began tracking her periods and was relieved when they continued to run in perfect twenty-eight-day increments. Her mother told her not to worry. "You've got plenty of time," she said. "You're only thirty-eight. I didn't go through menopause until I was fifty." But Hannah remembered what her doctor had said about her fertility ending way before that.

So when she saw an article in *People* magazine in July 2002 about a new technique to stop her eggs from aging, she was immediately intrigued. "This is a good idea," she told her mom and showed her the story of a twenty-seven-year-old woman who had lost an ovary to endometriosis and froze her eggs before the second one was removed. The author wrote that egg freezing was considered a great hope for "single women who want to delay childbearing until they find a mate." Although only seventy babies had been born around the world, Hannah was so encouraged by the concept that she was willing to try it. Her mom agreed to help her financially.

After visiting a fertility clinic in Seattle and learning that the procedure wasn't available there, Hannah called Dr. Kevin Winslow, one of the doctors mentioned in the article. He explained that she would need to complete her early monitoring appointments in Seattle, then fly to Florida for the retrieval. He laid out the disclaimers (risk of hyperstimulation, risks from anesthesia, risk of eggs not surviving or fertilizing) and quoted her a 30 percent success rate per transfer of embryos. If she was interested, she would need to act quickly. She was already thirty-eight and a half.

Hannah didn't know if 30 percent was considered good or bad odds. All she knew was that Winslow was offering a chance, and she wanted it.

She was used to traveling for work, so going cross-country to Florida wasn't a big deal. Her mom went with her, and they tried to make the best of the weeklong trip by shopping, dining at nice restaurants, and taking a side trip to Amelia Island.

In a departure from her usual style of downplaying her desires, Hannah had a rare confidence that the procedure would work. Winslow retrieved twenty-one eggs two months before her thirty-ninth birthday, and Hannah reveled in a profound sense of satisfaction. She wasn't passively waiting for her menstrual cycle to go haywire or waiting for a man to call. She was taking care of herself.

Sarah

Five months after attending the recruiting seminar sponsored by Reproductive Medicine Associates and Extend Fertility, I called the clinic for a consultation. I had looked up RMA on the Internet and learned it was ranked one of the best clinics in New York City. They had a good reputation and some experience making frozen-egg babies, which was rare in the field. I didn't know what else I could find out on my own.

I was assigned to Dr. Tanmoy Mukherjee, a well-spoken father of four, who repeated much of the information I had heard from the seminar. He added that even though the procedure was still considered experimental, he believed in it enough to recommend it to his own daughter. "We know it works," he said confidently.

But before I could freeze, I first had to see if I was eligible. Even at thirty-six, there was a small chance I was already infertile. On the third day of my next period, a nurse drew my blood to test my hormone levels.

I left the clinic pressing the bandage in the crook of my arm, unable to shake the heavy feeling of dread. I couldn't imagine learning that the game was already over just as I was getting my act together.

When the nurse called to inform me that my FSH level was 9.6 "Normal," she said—I did a little dance in my apartment and called Paul with the news. "That's good," he said. "I bet that's a relief." It seemed strange talking abstractly about something that could have such huge implications for us. It was like calling him with the news that our favorite wedding caterer would be free on Memorial Day weekend two years from now—you know, *just in case.*

The next step in the egg-freezing process was to send Extend Fertility a check for $13,055. According to my invoice, RMA received about $8,000 for medical care, and Extend received $5,000 for science and services, which included "access" to its "scientific knowledge," patented freezing solution, and trained clinic partners, as well as transportation of eggs to and from a storage facility, the first year of storage, and records management.

I signed releases stating that I was part of an ongoing trial, that I was aware my eggs might not survive freezing, and that it was unknown whether there was an increased risk of birth defects in children born from frozen eggs. I also understood that the risk of terrorist attack or natural disaster was outside the control of the storage facility. However, if an employee lost or damaged my egg, I would be reimbursed $400 per egg. I stopped reading. "That's it?" I exclaimed. I couldn't believe how much Extend undervalued my eggs. It was one thing for *me* not to appreciate the worth of my eggs by ignoring my biological clock; it was another thing for a *business* not to.

I also had to choose what I wanted to do with my eggs in the event that I died: Discard them or donate them to research? The consent form made it clear they would not be used to create embryos that could be destroyed during experiments. If I wanted to leave them to someone, I had to send Extend legal authorization. I didn't have any biological sisters, and I thought it would be creepy if my parents tried to make their own grandchildren with my eggs using a surrogate. (In 2011 an Israeli court allowed the family of a seventeen-year-old who died in a car crash to harvest her eggs.) Maybe someday I would leave them to a friend or family member who needed them. But for now, I checked that I would donate them to research.

I then met with the clinic's egg-freezing coordinator, and we went over the list of hormone injections and clinic requirements, including more blood tests to detect various forms of hepatitis and HIV. She told me to ask my ob-gyn for the results of my latest gonorrhea and Chlamydia tests. I would also have to attend an egg-freezing class to learn how to give myself the shots and schedule a mandatory $300 counseling session with Dr. Georgia Witkin, the psychologist who had spoken at the seminar. I assumed Dr. Witkin had to attest that I was emotionally fit to undergo the procedure, but I had no idea how an emotionally fit egg freezer was supposed to behave. Should I maintain a delicate balance between enthusiastic and somber? Hopeful and realistic? If I was *too* excited, would I be flagged for not fully appreciating the risks involved? I was relieved when she explained that the point of the session was to answer *my* questions.

She glanced at the questionnaire I had filled out beforehand and remarked that I fit the freezer profile almost perfectly: late thirties, highly educated, proactive in protecting her options. According to Extend's "motivational assessment" of women who chose to freeze their eggs, 75 percent had advanced degrees and described themselves as "intelligent" and "extroverted." I suddenly felt proud. I wasn't someone who had steered her life poorly; I was an overachiever!

Then Witkin got down to business, "So what brought you here? A birthday? A breakup?"

"Well, I know I want children," I explained. "But my boyfriend isn't sure, and in any case, I don't want kids for a couple of years. But I'm turning thirty-seven soon."

"How does your boyfriend feel about egg freezing?" she asked.

It felt like a trick question. "He's thrilled," I exclaimed. "What man wouldn't be? It takes the pressure off both of us. If this wasn't an option, I would have had to make some hard decisions a while ago."

Witkin said I was lucky that my boyfriend was so supportive. She had several clients whose partners felt threatened by egg freezing. One guy asked, "Well then, why do you need me?" Another feared that freezing meant his girlfriend was about to break up with him. Witkin repeated what she had said at the seminar: that the act of freezing made women feel as if they were in charge of their romantic and reproductive lives. It also made some women more open to having children on their own. Since their ability to have a baby no longer depended on a partner, they could separate procreation from romance. "There's something powerful about having those eggs in the freezer," she said.

I mulled that over for a few minutes. But I didn't want to separate procreation from romance, I thought. I wanted egg freezing to help my romance lead to baby-making, either by giving me and Paul more time to resolve our differences or by giving me the chance to find someone else. Egg freezing was supposed to help me avoid becoming a single mom. I could do that on my own right now, if I wanted. Or did egg freezing make these women feel so large and in charge that having babies by themselves suddenly seemed less overwhelming? The partner part just became another detail.

I looked at the list of questions I had brought. "I don't know how to make sense of the success rates," I said. Witkin repeated what I had been telling myself: the technology was good now, and it would be even better by the time I needed to use the eggs—*if* I needed to use them.

After my counseling session, I joined two other patients for class. We took careful notes on how to mix the medications that came in powder form and load them into a syringe. When the nurse saw me flinch at the size of the needle, she quickly reassured me that I was looking at the mixing needle; a smaller one was used for injecting. She insisted that the idea of giving ourselves shots seemed more ominous than the act itself really was, explaining that women who endured several IVF rounds became very comfortable with it. She instructed us not to drink, run, or engage in strenuous exercise so we wouldn't jiggle our swollen ovaries. And we probably wouldn't feel like having sex. We were also supposed to be on the lookout for ovarian hyperstimulation syndrome, which affects 1 percent of patients, and at its most extreme causes one's ovaries to burst. Symptoms include sudden weight gain, vomiting, nausea, diarrhea, dark urine, and shortness of breath.

When my payment cleared, my egg-freezing coordinator ordered my medications, which cost $3,000 on top of the $13,055 egg-freezing fee and $300 psych consult. I couldn't stand to look at my bank statements and see my obliterated savings account. I had wanted egg freezing so much that I hadn't thought much about the money. But it began to sink in that I had left myself financially vulnerable. I had traded my financial backup for a fertility backup.

The drugs arrived overnight in big impressive-looking foil bags and had to be rushed to the refrigerator. I organized my syringes, needles, alcohol pads, and vials in neat rows on my counter and taped notes to every mirror with my medication schedule: one shot in the morning and two more in the evening.

On the third day of my period, after a month of taking birth control pills to regulate my cycle and lower the risk of hyperstimulation, I went in for another blood test and an ovarian ultrasound that showed dozens of follicles containing tiny eggs that looked like chocolate chips on the monitor. They were waiting to be assaulted by mass amounts of manufactured hormones. Each month, my body naturally selected the best egg from the bunch to offer up for ovulation. The goal here was to make them grow all at once. Later that day, a nurse called and gave me the green light to start the shots that evening.

I like to think I'm a tough cookie when it comes to physical pain. I can take out my own stitches and fish for deep splinters. But even I was squeamish when I imagined sticking myself with needles to make my ovaries produce at least ten times their usual output.

I started my egg-producing regimen with an injection pen filled with Gonal-f, which is a synthetic version of FSH and made from genetically modified Chinese hamster ovary cells cultured in a lab. I dialed the dosage, screwed on the half-inch needle, and took several deep breaths to summon my nerve. The nurse had recommended injecting into my stomach an inch below my belly button. So I grabbed a roll of flesh, wiped it with alcohol, scrunched up my face, and jabbed the pen. I held it still, fumbling to push the top down to squirt the medicine inside and quickly pulled it out. Thankfully the needle was thin, and I could bear it.

The next shot was more complicated and involved mixing a powder called Repronex with saline solution, loading it into a syringe, and replacing the mixing needle with the injecting needle. This contained more FSH to make the eggs grow as well as luteinizing hormone to help them mature. The powder was made from the purified urine of postmenopausal women, a bizarre fact that made me wonder if they actually got paid to pee or did it for free, as their last procreative gift to the next generation. This shot burned a little, and I gritted my teeth while I pushed down the plunger. The next day the site had a sore rash that looked like a mosquito bite.

Since I am extremely sensitive to hormonal changes—to the point that my parents counted down my periods until I left for college—I had prepared for an onslaught of moodiness and warned Paul accordingly. However, I was surprised at how good I felt. I had a mild buzz and blew through my days being pleasant and prolific. Three days later, I marched into the clinic for my monitoring appointment, eager to learn how many eggs were growing.

The waiting room was full of patients anticipating their ultrasounds and blood work. The mood was somber, and I imagined all the private dramas taking place. Lesbian couples snuggled next to each other. Men sat looking bored or uncomfortable. One woman cried in the corner. Did she just receive devastating news, I wondered, or was she a wreck from the hormones? Others punched their BlackBerrys or badgered the receptionist about their being late for work meetings. One woman played loudly with her toddler, crooning, "We're going to see Daaaaddy soon." I was stunned as I watched the room watch her. Was this woman actually showing off her child? I learned that the clinic later posted a sign asking patients to make child care arrangements in advance due to the "sensitive nature of our practice."

During my ultrasound, I could barely stand the suspense as Dr. Mukherjee pushed the vaginal probe against my ovaries. The pressure was uncomfortable, but I didn't complain. I wanted him to find every last egg follicle, which had now grown to the size of dark cherries on the screen. "How many?" I asked as he pulled out the probe.

"I see six on this side and six on that one," he said, looking satisfied.

"That's it?" I exclaimed. I was thrilled the hormones were working but was disappointed at my potential harvest. Why hadn't I gotten one of the bumper crops I had heard other women getting? He explained that the number was normal for my age, but his reassurance provided little comfort. I had always felt advanced in womanly ways: I was the first to start my period in sixth grade and had breasts before everyone else.

Now I felt as if my egg count was somehow a diss on my femininity. Not to mention a horrible financial return.

I knew it was the quality, not the quantity, of the eggs that mattered, but I wondered if I had messed up somehow. Had I drunk too much wine at a dinner party? My doctor had said I could drink one serving of alcohol a night, but I hadn't paid close attention to how often the host had topped off my glass. Did I squirt out too much medication when I tested the needle? Did I do the injections at the right times?

Mentally I still felt like a young woman with her whole life ahead of her. But the hard truth was that, at almost thirty-seven, I had proof that my fertility was dwindling.

By the fifth day, my can-do attitude had given way to lethargy. I was surly, had a persistent headache, and could feel my ovaries churning. My arms were bruised from all the blood draws, and my stomach had grown so grotesque with bloat I couldn't fit into most of my pants. On the eighth day, I added another shot called Cetrotide, which prevents the pituitary gland from secreting large levels of luteinizing hormone. Otherwise, I would waste the eggs by releasing them into my fallopian tubes instead of holding them in my ovaries, so they could be removed during surgery.

It's normal for some follicles to stop growing or die off during stimulation, and I was now down to nine eggs. On the tenth day, I took my last shot: human chorionic gonadotropin, which goes by the brand name Ovidrel and is the trigger to mature the eggs and release them for collection. The shot must be given thirty-six hours before retrieval and timed precisely to catch the eggs before they headed for the tubes.

During my surgery, Dr. Mukherjee collected all nine eggs. That afternoon I surrendered to deep sleep at Paul's apartment and awoke to him kissing my face. "Dinner's ready, babe," he said as he opened his bedroom blinds to let in the last of the evening light. I felt a soothing stillness as I followed him and the smells of Italian takeout down the hallway.

What doctors don't tell you is that the real hormone misery begins a few days after your retrieval, as your body goes into a PMS conniption trying to have a period. The tension in my shoulders was intolerable.

Nothing—not mimosas, massages, medication, or meditation—made me feel better. I wanted to curl up on the couch with a six-pack of red velvet cupcakes and watch *Oprah* all afternoon.

A couple weeks later, an envelope from Reproductive Medicine Associates arrived containing a photo of my frozen eggs and a breakdown of my egg count. Out of nine eggs, one was abnormal and discarded. Eight were frozen: six mature eggs and two immature eggs. (The latter were considered long shots but kept anyway in case the egg-maturing technology improved by the time I was ready to thaw them.) I stared in disbelief at the photo of six shiny discs floating in a chemical solution. That was my final count.

Dr. Mukherjee had estimated it would take twelve to fourteen eggs at my age to have a "very good" chance of success. If all six eggs survived freezing and thawing, I might end up with a few embryos. According to averages determined by the Centers for Disease Control of all the IVF cycles in the country, I had about a 30 percent chance of becoming pregnant and delivering a baby using *fresh* eggs. The fact that I went to a good clinic meant my chances would be higher, but that didn't include the handicap for using frozen eggs.

For the first few weeks following the retrieval, I gave myself permission not to think about Paul, fertility, or eggs. Then I took stock. I waited for my head to flood with anxiety and my shoulders to stiffen, but I remained calm. The familiar panic didn't exist any more. "You just need to find a way to get more eggs," I told myself. "You can do that."

Dr. Witkin was right. There *was* something empowering about having those eggs in the freezer.

4. *Men*

Monica

Egg freezing had made Monica feel more relaxed about her future, but the procedure couldn't soften the blunt reality of her fertility math: even if she met someone at age thirty-seven, she'd still be pushing forty by the time she had her first child. She'd barely have time for a second. She had thought egg freezing would ease some of the pressure to date, but if anything, it made her painfully aware of how little time she had left.

But in the months after freezing, Monica couldn't seem to slow her life down to make dating a priority. As a project manager for an electronics company, her travel schedule that fall was especially heavy, and she was getting fed up with the online dating site eHarmony. The volume of computer-generated matches had dwindled, and she wasn't impressed with the men she did manage to meet. "Screw this," she said in exasperation one evening. "There's got to be a better way!"

By the holidays, she had resolved to revamp her whole approach. Nearly eight months had passed since her egg retrieval, and she was no closer to her goal of falling in love and starting a family. The chest pains never let up either. She had to be strategic about protecting her precious free time for men who were serious about her. To weed out the wishy-washy, she decided to tell dates up-front that she was looking to settle down and have children. The approach was controversial within the dating advice industry. While some experts believed such candor was necessary to make sure two people shared the same goals, others warned it might give men the impression that the woman was more interested in looking for a "baby daddy" than a relationship. Monica believed that the type of man she was looking for wouldn't be scared off by the subject.

She joined Yahoo! Personals, an online dating site with a large membership. In her profile, she described herself as playful, independent, athletic, intelligent, funny, and confident and chose the headline "Sporty travel maven seeks mate." Hoping to attract only sincere replies, she wrote, "I'm serious about finding a lifelong partner with whom I can share my interests, goofy sense of humor, travel experiences, and good conversation." She checked the box indicating that she wanted kids.

She received several dozen emails on the first day and a steady stream thereafter. More than three-quarters were under forty and wanted kids too. Monica wondered if guys were clueless about the biological clock. Otherwise it didn't make sense for them to contact a thirty-seven-year-old woman. Maybe they just liked her profile photo and wanted a date and weren't thinking that far ahead. Whatever the case, she planned to tell them about her frozen eggs within a few dates.

Soon Monica was going on several dates a week. She loved having her pick of men, but she quickly realized she was in over her head when she was dating three guys whose names started with "S" and she couldn't keep track of them. She almost embarrassed herself when she accepted Sean's invitation to watch *The Sopranos* and called out "Scott, I'm here!" when she arrived at his house. Thankfully he didn't hear her. Another time, she asked a date about his sister's surgery and was reminded that it was his brother who had been in the hospital.

To avoid further mix-ups, Monica made an Excel spreadsheet of every man in play. Next to a thumbnail photo she copied from Yahoo! Personals, she listed each prospect's age, education level, number of family members, hometown, what they had talked about on dates, personality traits she liked, and whether she owed him an email or a phone call. She organized the information on one page and carried around her fact sheet so she could reference it during unexpected calls.

"That's classic Monica," her friends teased her about her business-like approach to love. Some warned her not to juggle too many men at a time or else she would burn out and have a hard time connecting with one. She knew she was being ambitious, but she loved the sense of control. She was done wasting time waiting for love and was simply managing her pipeline as she would a project for work. Dating should have been as important as this all along.

In March 2006, less than one month after joining Yahoo! Personals, an email with the subject line "I've been waiting for a sporty travel maven" showed up in her inbox. Monica clicked on the sender's profile and was struck by the black-and-white photo of a thirty-seven-year-old arty-looking guy with shoulder-length hair and a well-tended beard who worked at a theater company. She usually went for conservative corporate types, but she liked this man's big genuine smile and sense of humor. "Hell, why not try the arts?" she thought.

They traded emails, and Monica learned his name was Adam; she admired his eloquent, clever writing style. On the phone they had an instant rapport, and he entertained her with silly voices. Other characteristics impressed her: he had followed his dreams by making a good management career out of his love for theater, was close to his family, and showed interest in her life and job—a nice change from the self-absorbed talk of her other dates. By the third phone conversation, she agreed to meet him.

Both lived in New Jersey about an hour from New York City, she in Basking Ridge in the middle of the state in Somerset County, and he in Red Bank in Monmouth County near the Shore. So they planned to meet for Sunday brunch in Manhattan. He took the train and met her bus at Port Authority. Monica usually didn't travel so far for a first date, but she had a good feeling about Adam. When she spotted him in the terminal, she suddenly felt nervous. He was just as cute in person as he looked in his pictures. His eyes lit up as he looked her over, and she could tell he was attracted to her too.

They chatted nervously as they walked to an Italian restaurant Adam had picked out in the Theater District. After the meal, he suggested going to Central Park. It was a gorgeous spring afternoon, and they strolled down Poet's Walk. When they sat on a bench near the Seventy-second Street band shell to listen to a saxophone player, Adam leaned over and kissed her. It was the first time in years she had wanted to kiss someone back.

By the evening they were walking around the city holding hands and stopping every few blocks to give each other quick pecks. Occasionally he pulled her into a doorway and kissed her more passionately. Their first date lasted ten hours, and she had to sprint to make the last bus home.

Monica usually held back her feelings until she had been on several dates with someone, but later that night she sent Adam an email: "There is something about you that is so refreshing, and it gives me a tingle inside just thinking about you. A good tingle, mind you, and one that I have not experienced in quite some time." She signed the note, "Un *beso*, Monica."

She was giddy when she read his response the next day: "I felt your sweet *besos* on my lips all night long on the trip back up north, through the night, and this morning when I awoke. And if I bury my face into my coat and breathe deeply, I can still smell the faint traces of your perfume, and our day comes rushing right back into the present. . . . It felt like we already knew each other but with the fresh surge of excitement of something just starting."

Within two weeks they decided to take down their online profiles. Within three weeks they had met each other's families. Within three months Adam asked her to move in.

When Adam brought up the idea of living together, Monica thought it seemed the obvious next step in their relationship. They lived an hour apart, and both were getting tired of the long drive. Since he had to be close to his theater and often worked evenings, it made sense for her to move in with him. She found a tenant for the townhouse she owned and moved her belongings to Adam's 1920s yellow bungalow. With its green shutters and two big flowerpots on the porch, the house looked as if it was out of a storybook. They planned to buy a new house together some time during the next year, so she didn't try to redecorate. But she hung up her favorite artwork, propped up framed photos of the two of them, and moved in her bed and throw pillows to feel at home.

Monica loved living with Adam. They would read together every night in bed or in front of the living-room fire on chilly fall Sundays, lying at opposite ends of the couch so he could massage her feet. He liked to cook and would often make roasted red pepper soup or grilled chicken fajitas. While he was in the kitchen, she'd sit on a barstool, pour a glass of wine, and tell him about her day.

Monica felt closer to Adam than to any of the other men she had ever dated. She marveled at how they communicated so openly and respectfully. They both told each other how much they loved the other, whereas in her prior relationships it had seemed as if one person always had the upper hand.

On their second date, Monica had told Adam she wanted a family and had frozen her eggs as an insurance policy. "That was smart," he responded, admitting that he had been thinking about becoming a dad for the past couple of years. They didn't talk much about children for several months after that, until a Labor Day barbecue they attended to celebrate the birth of her nephew. Monica beamed as she watched Adam coo at the sleeping infant he called "Shorty." She became choked up imagining him playing with a baby of their own.

A few days later he announced, "Shorty sealed the deal with me. Let's start trying to get pregnant." At first she was speechless. They weren't even engaged. They had talked theoretically about having a baby, but she hadn't intended to start a family so soon. "Why wait?" Adam pressed. They were in love, he said, and they were both getting up in years. Monica was now thirty-eight, and he was only six months behind her.

At first Monica was alarmed at how fast Adam was moving, but she didn't feel that lingering doubt that made her question whether they were rushing a major decision. Having a baby with him felt right. She couldn't imagine a better endorsement of their relationship. They continued having sex nearly every day, now without condoms.

She didn't have to wait long before all the other details fell into place. One night before Thanksgiving, Adam came home carrying a cake and a bottle of Champagne. She noticed his eyes were watering; there was something sunken in the white sugar frosting. "I love you, and I want to spend my life with you," he proclaimed, setting the tray down in front of her. "This is just a placeholder," he said of the modest diamond in the cake. "Will you marry me?"

"Oh my God!" Monica shouted. "Yes!"

Kelly

Nine months had passed since Kelly had broken up with Steve, yet she still hadn't started dating again. She didn't trust her motives. Shortly after the split, she frequently longed for a man's attention and company, and she feared rushing into another bad relationship to alleviate her loneliness. She realized that except for a few months here or there, she had been romantically involved with someone ever since high school. She concluded that she needed to learn how to be alone. Her eggs could wait in the freezer.

She wanted to develop confidence in her judgment to pick a good partner. Even if she didn't end up marrying the wrong guy, she didn't want to waste several more years learning he wasn't right for her. At thirty-nine, she didn't have time for many more rounds of trial and error. She had to sharpen her ability to quickly end relationships that weren't working. Her logic: taking a break from dating was actually a better strategy to help her make progress toward her goals.

The phrase *taking a break from dating* is used so commonly by women that it's become a cliché. The main premise and promise is that by taking time off to focus on yourself—to improve your appearance, learn relationship skills, practice yoga, change jobs, or undergo psychotherapy—you will emerge as a more desirable dating candidate. For some, it offers the chance to reset one's romantic life, to heal from previous relationships, regroup, and renew the motivation to date. For others, though, it can be a trap, since it's tempting to want to improve oneself indefinitely, either by striving for an unrealistic state of perfection or using the time off as an excuse to avoid the hassle and rejection of dating altogether.

Still, Kelly was behaving the way critics of egg freezing feared she would. Had the frozen eggs dulled the urgency to get her life in order to start a family? It was bad enough she had to freeze her eggs because she hadn't heeded the clock in the first place. Now she would get even further off track.

During her break, Kelly was surprised that she was rarely alone. She attended family fish fries and children's birthday parties. She joined the YMCA's volleyball and basketball leagues. She helped put on annual banquets for the Chamber of Commerce and oyster roasts for her work. She still sang in the church choir (now without Steve, who resumed going to the earlier service). She took guitar lessons and exercised every evening with her running group.

She also got used to going to the movies or park by herself on holidays and Sundays, days she had formerly dreaded because they seemed reserved for couples or families. She wished it hadn't taken until she reached her late thirties to learn how strong and resilient she really was. She had been focused on men for so much of her life that she had never really known how much she could enjoy her own company.

The dating platitude turned out to be true: when she realized she could be happy without a man, she found herself ready to welcome one into her life.

Six months after freezing her eggs, and more than a year after her breakup, Kelly spread word that she was ready to date. Old ladies from her church set her up with their sons. Her friend Jill, whom she met at a spiritual retreat, volunteered to help find her a husband. She met men through friends and networking events. Some were in their fifties and seemed too old; others were scared off when she told them that she wanted kids.

She dated a guy long-distance from Tampa for ten months but broke up with him when it became clear he wasn't over his divorce. She had another months-long relationship with a friend that turned romantic but quickly fizzled. She cut others loose more quickly: a father of five who lived in a different state from his kids and a guy who bragged that he bought only Prada shoes. Her family joked that she was becoming so good at getting rid of guys that it was an accomplishment for any to score a second date.

Sometimes Jill, who was married with two kids, would call with a suggestion of someone she should meet. Usually Kelly was open to blind dates, but after suffering through several disappointing ones, she followed the advice she had read in a relationship advice column to make a list of her requirements. She had been hesitant to limit her search in case she missed someone who was a real catch but fell just outside her criteria. But she decided to come up with a basic list: she wanted her future husband to be under forty-five, Christian, and health-conscious. He should want kids (more if he already had some) and plan on being an involved parent, not a detached breadwinner who left the daily details to her. And she wasn't interested in couch potatoes or workaholics. When Jill confessed that her setup was in his fifties and didn't want any more kids, Kelly promptly vetoed him. "I'll keep looking," Jill promised.

Kelly started to wonder if she had become so good at detecting red flags that she hadn't given some guys a fair chance. Despite her calm and confident image, she often woke up at 5 a.m. awash in anxiety. "What if I'm single forever?" she asked herself. But she told herself that she now knew how to be happy.

Part of her even enjoyed not being tied down anywhere or to anyone. Maybe she could go back to school and study interior design or sell her house and travel around the world. She toyed with the idea of moving to Italy or to Atlanta to be near her sister. Who knew when or where she would meet someone?

"When love happens, it happens," she told herself whenever she was upset. She remembered a prayer she had read, about not being so paralyzed with worry about the future that she missed out on the moment and season of life. It was easier to take the message to heart, though, when her eggs were frozen and her season was still in swing.

For extra inspiration, she tacked to her refrigerator a list she found of older celebrity moms, including Holly Hunter, who gave birth to twins at forty-seven, Jane Kaczmarek, who had her third child at forty-six, and Jennifer Beals and Ming-Na, who gave birth to sons at forty-one.

She had learned from Dr. Winslow that some probably had used donor eggs, but as she grew older, she didn't think that detail mattered so much anymore. They were still moms.

Kelly had been back on the dating market for nearly two years when Jill called again. "I've got someone for you," she announced. That someone was her recently divorced boss, Dan. He was forty-six and the dean of the local college where Jill worked. "I'd have job security forever," Jill joked. Dan lived in Brevard, a small town two hours west of Charlotte but just a few miles from where Kelly had grown up and where her mom and three brothers still lived. Best of all, Dan had eight-year-old twin boys and wanted more kids. But Kelly was wary that he had been single for only a couple of months. She definitely didn't want to be his rebound relationship.

Still, she told Jill to give him her phone number, and Dan called a few days later. The conversation flowed easily, and she admired the loving way he described his sons. After talking a half-dozen more times over the next two months, he suggested they meet. But she didn't want him to spend two hours driving to Charlotte before she even knew if she liked him romantically. When she mentioned that she'd be traveling near his town on her way to meet her brother and niece for a hiking trip in the Smoky Mountains, he asked, "When you're in the area would you give me the privilege of taking you to dinner?"

"Certainly," Kelly replied enthusiastically. "I would love to."

There weren't a lot of places to eat in his small town without drawing attention, so Dan suggested, "Why don't I just cook you dinner? Are you comfortable coming to my home?" Normally Kelly would never go to someone's home on a first date, but she had a good feeling about him. So on a warm April evening, she put on her favorite vintage white linen camisole and finest pearl jewelry and rang the brass doorbell of his immaculate Victorian home.

At first she wasn't sure how she felt about Dan. While she liked his clean-cut style and positive, easygoing way, she wasn't immediately attracted to his bushy salt-and-pepper hair and tall lanky build. But she was touched by all the effort he had made in cooking her dinner. He proudly served her a rib roast with gorgonzola cheese sauce, red potatoes, and salad, plus cheesecake for dessert. When he apologized for his brand of pinot noir and said he had struck out finding his favorite label, La Crema, Kelly was delighted at his good taste. She had tried La Crema a few months before, and it was her new favorite wine. She couldn't find it anywhere either.

Over dinner they talked about their divorces, bad dates, and life paths. Kelly was increasingly impressed with the profile of Dan that was emerging: he was a doting father and a smart, respected figure at his college and church. Since having a heart attack at age forty-two, he had become vigilant about his health and exercised regularly. But when Kelly learned that he had undergone a vasectomy, her heart skipped a beat. She wasn't worried about his ability to have children, since she knew vasectomies could be reversed. She was alarmed that a man who said he wanted more kids would have taken such drastic measures to make sure he didn't have them. Dan explained that his ex-wife was so adamant about not wanting more children that he agreed to the surgery. At the time he thought he would be married forever.

Kelly thought it was a good time to bring up her fertility, since she wondered if Dan doubted her forty-one-year-old body's ability to have a baby. She knew that Jill had already told him that she had frozen her eggs to prove she was serious about having a family. Still, she wanted to make sure Dan understood that she wasn't in a mad rush to find a father for her children. "It took the pressure off me so I could find the right guy," she said during dinner. "That way, having a baby isn't constantly on my mind."

Kelly realized these weren't usual conversations for a first date, but she felt they had gotten to know each other over the phone and was relieved to get the big questions out of the way from the start. She liked Dan's answers. He confessed that he didn't know much about egg freezing but thought it was a smart idea. He also explained why he had started dating so soon after his divorce. He wasn't mourning his ex-wife, he said. He missed being part of a family.

Hannah

As part of her psychotherapy, Hannah had to reexamine her thoughts about romantic relationships, especially her knee-jerk tendency to interpret men's rejection as proof she was unlovable. She knew she had to keep going on dates. "The more, the better," her therapist added. So she summoned her courage and joined an expensive local matchmaking service in Pike's Place in downtown Seattle, which she hoped would lead to better quality matches than joining an Internet dating site. She spent $1,000 to meet ten men over the course of a year but liked only one. They went out a few times, and he stopped calling. She tried to brush it off.

Hannah also became more open to dating men who already had kids. She and Lori liked to joke that since they had missed the first marriage cut, they would be first in line for the "second rounders," men coming off their first divorces. "At least they can commit," they'd tell each other. The two pals had been nearly inseparable since college, and friends referred to them as "the Hannah and Lori Show." Lori, an animated, attractive blonde who could easily chat up men, had a social ease Hannah envied.

Hannah had always assumed Lori would get married and become a mother, just as she assumed she would have by now. But at nearly forty, the two women were still very much single. Although each hoped the other would find love, they shared a certain comfort in the fact that they weren't alone. But after freezing her eggs, Hannah noticed a shift in their friendship. She felt guilty that she had an advantage Lori didn't. Her friend wanted children too but never mentioned freezing, and Hannah didn't want to push her. So they avoided the subject and focused on their usual pursuit: meeting men.

Hannah and Lori were at a beer garden where their college alumni group had gathered to watch a televised Washington State football away game when one of their former sorority sisters dropped a name that made Hannah's stomach flip-flop. "Nate Williams is here," she announced, adding, "and he's divorced." Hannah was curious to know why Nate, whom she hadn't seen in nearly twenty years, had suddenly shown up at an alumni event. Lori's ears perked up at the name too because she remembered Hannah had dated him briefly in college. "You and Nate!" Lori exclaimed. "We should find him. Let's see what he looks like now."

Hannah and Nate hadn't dated in the traditional sense of the word. Rather, over the course of a couple of months, they attended a few fraternity dances together and made out afterward. Hannah, a sophomore, suspected that Nate, a junior, was looking for a girlfriend. She didn't want to be tied down to one person at age nineteen and declined his next invitation. Soon afterward he started dating a woman named Amy who would become his wife. Although Hannah hadn't wanted Nate for herself when she had the chance, she was somewhat indignant that she had been so easily replaced and secretly pined for him until graduation. Over the years she had thought about him from time to time.

As soon as Lori heard Nate's name, she quickly went into action. Within minutes she had located him, positioned Hannah in front of him, and conveniently started talking to his neighbor. Hannah was inspired by the way Nate's eyes lit up when he recognized her. "How have you been?" he asked. "I haven't seen you in a long time."

"Good to see you," she said breezily, feeling uncharacteristically brave after a couple of beers. "Sorry to hear about you and Amy." Time had served Nate well, and she found him much more attractive than before. He had grown into his gangly body, and she admired his broad shoulders and slender physique. His boyish face had become chiseled and weathered, and he wore a sly smile. She also felt attractive in her tight jeans and fitted alumni sweater that showed off her petite figure.

They quickly exchanged updates: Nate was a civil engineer, a father of two, and had been divorced for about three years. She told him about her job as a fashion designer and how she had recently sold her condo and bought a house in the Greenwood neighborhood of Seattle. She was self-conscious about admitting that she had never been married, though. "I sort of had some crummy relationships," she explained.

As soon as Hannah sensed a lull in the conversation, she mentioned she should find Lori, who was scouting out another corner of the patio. "Well, it was great seeing you!" she said as she kissed Nate's cheek. "Why don't you give me a call? I'm in the book."

A couple of days later, Nate did call. He asked her to dinner at a nice seafood restaurant that overlooked Lake Union, and over a bottle of wine they joked that it was a step up from drinking "Derailers," their favorite college cocktail, consisting of rum and pineapple juice served in a large bucket with a fistful of long straws. They reminisced about college and old friends, and Hannah felt good to have a shared history. He told her more about his divorce and the shared custody of his eleven-year-old daughter and nine-year-old son. At the end of the meal, he walked her to her car and gave her a goodnight kiss. His technique had noticeably improved since college.

"Wow," she thought to herself as she drove home. "This is a really nice guy. What was I thinking?" She remembered he had been nice back in college too. But back then she had imagined herself to be hip and worldly and fantasized about meeting hip and worldly men. She thought musicians with long hair and goatees were more her style. After all, she wanted to go into fashion.

But at this point in her life, she loved the idea of *nice*.

Sarah

When an egg is frozen, it sounds like French fries plunged into hot oil as an embryologist dips the holder into smoky liquid nitrogen. Fifteen minutes earlier, a freezing solution called a cryoprotectant shrivels the egg like a raisin as it draws out the water, then plumps it up by working its way inside through osmosis. The egg will wait indefinitely in a storage tank until the patient is ready to use it.

When an egg is thawed (or warmed, depending on the lingo), it is placed into a series of sucrose solutions to coax out the cryoprotectants and let the water back in. Ten minutes later, it's ready to fertilize.

Jeffrey Boldt let me observe his magic in his lab at Community Hospital in Indianapolis one fall afternoon. He is an embryologist who works with Donald Cline, a fertility doctor; the team is one of the leaders in babies born from frozen eggs in the United States. They started their program in 1997, when Cline asked Boldt, an expert in the fertilization of sea urchin and mouse eggs, to join his practice.

This particular morning, Boldt was working on the eggs of a thirty-five-year-old IVF patient. His assistant wheeled in an incubator containing the windfall from the morning's retrieval: seventeen shiny eggs. Boldt looked through a microscope and determined that ten eggs were of good quality, meaning that the fluid (or cytoplasm) wasn't dark or granular. He set aside half to be fertilized right away and froze the other half. Although Boldt had earned his stripes with his slow-freezing method, he has since switched to vitrification, a flash-freezing technique that consistently results in a 90 percent egg survival rate. Vitrification cools the eggs so quickly that ice never gets a chance to form, so there is a lower risk of chromosomal abnormalities. During slow freezing, the cold temperatures force the fragile barrel-shaped spindle that holds the chromosomes to separate. It comes back together during thawing; however, there is always a risk that the chromosomes won't realign correctly. But vitrification puts the egg into suspended animation, so the spindle doesn't move at all.

Boldt got to work fertilizing the remaining five eggs by inserting a single sperm into each, using the ICSI technique. Maneuvering micromanipulation joysticks, he lifted a sperm with a suctioning probe that held it still. Then he grabbed an egg and slipped the sperm under its shell, or zona. Next he pierced another internal layer to place the sperm in the egg's core. The zona gave way easily, but the core resisted the needle and buckled inward, like a balloon being poked with a chopstick, until it finally accepted the sperm. Then it closed up its puncture wound to shut out other sperm. I couldn't believe I had just witnessed the moment of conception. It all seemed so routine: life created in a clean white lab with a Wendy's commercial playing on the radio in the background.

Unlike other practitioners who had developed egg freezing to help Clock Tickers, cancer patients, or donor egg recipients, Dr. Donald Cline was attracted to egg freezing because he didn't like freezing embryos. In the early 1980s, when embryo freezing was discovered, it was common practice for doctors to fertilize all the eggs an IVF patient produced. They didn't want to transfer all the embryos for fear of creating triplets, quadruplets, or more, so they froze the extras. Patients who didn't become pregnant the first time around or who wanted more children later returned to thaw their frozen embryos. However, Cline, a devout Christian who advertised himself as a "pro-life" fertility doctor, became troubled when some patients left their embryos in storage indefinitely or let them be discarded. "If you do nothing, an embryo divides. All you have to do is keep it warm," he explained to me at his small practice.

So Cline stopped freezing embryos. But he was uncomfortable throwing out the rest of a patient's eggs that weren't used on the spot, especially if a couple could afford only one IVF cycle. In 1997 he became inspired by the Italian research and hired Boldt. After years of trial and error, they thawed the eggs of a thirty-two-year-old woman whose fresh IVF cycle had failed. They injected four of her thawed eggs with her husband's fresh sperm and transferred two embryos. In July 2001 a healthy little girl named Emma was born.

Cline started offering egg freezing to patients in 2004 and was one of the few doctors who accepted those whom he fondly called "the ticker ladies." He said he felt compassionate toward women worried about running out of time because he had seen how much his patients had suffered when they couldn't have a baby. "Infertility is an absolute crisis, especially for women. Little boys fantasize about being a policeman or a baseball player. Little girls fantasize about being a bride and a mom. Ask a six-year-old if she wants to be a mom, and she'll say 'Yes.' If you ask a boy if he wants to be a dad, he says, 'Huh?' If, a few years after being married, she can't get pregnant, it's a threat. To help them through it is very satisfying."

There was a hitch, though: Cline would thaw eggs only for married women. No single women. No cohabitating partners. No lesbians. He said he would ship the eggs to another doctor, but he had the right to choose which patients he treated, and he thought children fared best within a traditional marriage.

I was struck that a pioneer of such an empowering advance for women would place restrictions on who could ultimately use it. His approach sent a mixed message: he championed women who were out of sync with the mainstream by giving them a second chance at motherhood, but he punished them if they hadn't joined the mainstream by the time they were ready to use their frozen eggs to become pregnant.

I couldn't imagine feeling rejected by your fertility doctor because you didn't—or wouldn't—make it to the altar, or because your soul mate happened to be a woman. The fact that the link between being a bride and being a parent is weaker now than ever, with 40 percent of children born to unwed mothers, makes Cline's approach seem particularly out of touch. One study of 240 prospective egg freezers at a New York City fertility clinic found that nearly 60 percent said they would consider single motherhood if they hadn't found a partner. What's *ideal* in Cline's opinion isn't what's happening in modern-day America, and it seemed patently unfair to deny the mothers of nearly half the nation's new children access to what would soon become standard reproductive science.

As I left Cline's office, I tried to remember when I had first wanted to be a mom. I don't remember wondering whether I *would* be one; it was just what girls did. As a preschooler, I toted around my Rub-a-Dub Dolly and Baby Alive, who mechanically moved her mouth so you could "feed" her a gelatin mixture that exited onto a doll diaper. She lasted until I fed her cottage cheese, which rotted in her stomach. I was the oldest of four children, and I knew all about taking care of babies. As a teenager, I was the most popular babysitter in our neighborhood; my high school friends joked that I'd be the first to become a mom. In college at the University of California at Berkeley, my girlfriends and I drank café mochas on the steps of Sproul Plaza as we imagined our children running around together and traded lists of favorite baby names. When I backpacked in Europe, I bought a poster of beautiful drawings of farm animals and carried it home for my future nursery.

Then sometime in my late twenties, the dream just stopped. I was overwhelmed with layoffs, daily four-hour round-trip commutes to New Jersey, grad school, and beleaguered relationships. Where did a baby fit in there?

After my visit with Cline and Boldt, I drove around suburban Indianapolis, which was decorated for Halloween. As I admired the plastic pumpkins on the porches, tissue-paper ghosts in pine trees, and cotton spider webs stretched around bushes, I felt a strange mix of nostalgia and disorientation, as when you see the same set of dishes from your childhood in someone else's home. I was supposed to be in one of those houses by now, making taco night dinner with my husband and kids and chatting about what we were all going to be for Halloween. I felt a deep burning in my gut and tightness in my throat and thought, "This is what regret feels like."

Monica

In December 2006, thirty-eight-year-old Monica was finally engaged and trying to get pregnant. Unlike her prior engagement, though, she was excited about planning her wedding. However, she wished she hadn't set the date for the end of June, which was only seven months away. Within a couple of weeks, her cousins were hounding her for details: Had she chosen bridesmaids' dresses? Interviewed caterers? Registered for gifts? It didn't help that Adam was preoccupied about an upcoming theater performance and had started moping around the house. She didn't know how to behave around him and often felt as if she were walking on egg-shells. She didn't think it was a good time to ask why he thought she wasn't pregnant after four months of trying.

The tension continued to build. Adam was distant for several days in a row at her parents' shore house over the New Year's holiday. Every time she asked him what was wrong, he just complained he was tired. She became so fed up that she questioned if they were really meant for each other. "What am I doing with a guy like this?" she asked herself, trying to contain the swelling anger she feared would push him further away. Mostly, though, she felt sick to her stomach. She was terrified he was getting ready to dump her.

Back home a few days later, Adam finally revealed what was bothering him. He was troubled that his brother's marriage was falling apart just a year after his wedding. His brother had realized there were red flags but ignored them. "How do *we* know we're truly compatible?" he pressed Monica. "How well do we really know each other?" He pointed out that they had met only eight months earlier, and he already hated the way she snapped at him when she was under a lot of stress. "When you do that, Monica, I don't feel close to you," he explained. On top of that, he resented all the sudden wedding pressure and wanted them to work through these problems on their own timeline. "Let's stop planning the wedding," he said. "We're not ready."

Monica was so relieved he didn't want to break up with her that the canceled wedding seemed like a mere detail. She admitted that she too didn't feel connected to him when he was depressed. They talked about how they had tried to fit too many major life events into such a short time frame and needed to make up for what they had missed during the past months, such as learning how to communicate and working out the kinks of living together. As a start, they both agreed to go into individual therapy.

Their sessions helped them sort through their own issues, but they were at a loss to deal with each other's, especially their different fighting styles. Monica had grown up in a loud and emotional family and was used to confrontation, whereas Adam's family was more reserved. The combination was a disaster. When Monica lashed out, Adam recoiled. He kept her at arm's length, and she felt helpless to repair the rift.

After several rounds, Adam suggested they see a couples counselor. "We can't do this on our own," he urged. Monica's therapist, Roni, agreed to see them every other week.

After a few sessions, they made a list of house rules to prevent and handle arguments: They shouldn't go to bed or leave the house angry. If an issue seemed hard to resolve, they should drop it and then revisit it after a few days, when they had more insight. If Monica came home in a bad mood, she should say, "I'm really tired. Something at work pissed me off. Give me time to unwind," instead of snapping at him. For Adam's part, he had to tell Monica what was bothering him rather than withdraw. For example, he should say, "What you're getting from me is my frustration. I'm not feeling appreciated." To help Monica better handle her stress, Roni recommended she start a morning exercise program and channel her feelings in a journal.

At first Monica was high on Operation Solidify Relationship. She and Adam both followed the rules, and their relationship improved dramatically. Even though she still felt hurt about the canceled wedding, and had been embarrassed when explaining the situation to her family and friends, she was moved by how committed Adam was to therapy and was grateful they had given themselves the chance to learn these skills before getting married. They were trying to avoid repeating the mistakes of others: not only was Adam's brother getting divorced, but so was Monica's thirty-five-year-old cousin, who had been married to her longtime boyfriend for just a year and had two young children. Monica and Adam were determined to do things differently. "Have you ever worked so hard at something?" she asked him one night after an evening of conflict resolution.

"No," he responded, nuzzling her neck in bed. "But it's totally worth it."

Even though the wedding was off, they hadn't resumed using birth control. Still, every month, Monica's period came and washed away her unfertilized egg. The periods also brought more chest pains; one episode lasted an agonizing five minutes. Should she see a fertility doctor? She had read that infertility was defined as the inability to conceive after a year of unprotected sex and that women over thirty-five should seek help if they didn't conceive after six months. She and Adam were almost at that point, and she was starting to panic. She scolded herself for becoming so upset. "Oh God!" she thought. "Maybe my stress is why I'm not getting pregnant." Then she reminded herself that even *fertile* women had only a 20 percent chance of becoming pregnant during a handful of days around ovulation each month.

Monica's cycles ran like clockwork, down to the day. So when she was two days late, she thought they might have finally hit the jackpot. She didn't want to tell Adam until she was sure, so for the next few days she reveled in the secret that his baby might be growing inside her. Every time she got dressed, she touched her breasts and felt relieved they weren't tender. For Monica, sensitive breasts were a sign that her period was imminent.

By the fifth day, she was over the moon.

Even if Adam was hesitant about getting married, Monica believed he still wanted a baby. She thought that if they could focus on a child, they could forget about their incompatibilities. A baby would tie them together and anchor a family.

On the sixth day, she saw a smear of brown blood and burst into tears.

"What's wrong, Mon?" Adam quickly asked when she emerged from the bathroom.

"I thought I might be pregnant," she stammered between sobs. "But I just got my period!"

"Don't worry," Adam said, awkwardly embracing her. "We have our eggs."

She was touched by his sweetness, but his hope made her anguish worse. If her frozen eggs didn't work, she might never be able to give him a baby.

Six months after calling off the wedding, Adam had yet to suggest setting another date. Monica took comfort in the fact that he never used the word *cancel*, yet she wondered whether she should still plan to sell her townhouse. They both had been so busy with work that they had stopped house-hunting.

Whenever she was anxious, she reminded herself that she had known Adam for only a year and tried to relax. Still, when she learned that her forty-year-old coworker was married within eight months of meeting someone, she felt a jolt of envy. When the woman sent Monica an email announcing she was pregnant a few months later, Monica felt her face burn. "Why isn't it working out like that for me?" she lamented, reminding herself to be happy for her friend.

In Monica's individual therapy, Roni asked the obvious question: "Why are you trying to get pregnant now if you're still working on the relationship?"

Monica was taken aback, but after a minute she told the therapist that getting pregnant and getting married were two different matters. "Marriage can wait," she said. "I'm on the brink of forty. Who knows how long it will take me to get pregnant?" She insisted that although

Adam had freaked out about planning a big wedding, they both still wanted a baby. They even teased each other about their parenting styles, speculating that Monica would be the disciplinarian and Adam the pushover. Monica swooned when she heard that kind of talk; Adam wouldn't indulge in it if he was uncomfortable with the subject.

Roni advised her to stay present in the relationship and stop fretting so much about the future. "Monica, you're focusing on your age and this timeline," she said. "What I'm sensing is that you're creating a lot of stress for yourself." Monica knew Roni was right, but sometimes she didn't want to work through her worry. She thought it was the kind of soul-shaking angst that would propel her into action and keep her focused on her goal.

She wondered why egg freezing hadn't quieted that frantic voice. It wasn't supposed to matter that she would be turning thirty-nine that summer. Why, then, was she reluctant to call off the baby race? "Why *am* I trying to get pregnant?" she asked herself in a moment of clarity on her commute to work one morning. "This guy doesn't want to marry me. What would happen if I *did* get pregnant?"

Later that night she brought up the subject with Adam. "I don't think we should be trying to get pregnant if we don't know if things are going to work out between us," she said matter-of-factly.

"Okay," he quickly responded. They agreed to abstain from sex around the time she was ovulating.

Adam's spineless "okay" landed with a thud. Monica had hoped he would have begged her to keep trying. The way she saw it, as long as Adam wanted a baby, he still wanted her.

By the summer they were exhausted—from work, from the rules, from the conflict, and from the conflict resolution. For twelve blissful days on vacation in Vancouver that August, they managed to just enjoy each other. They had sex constantly, and Monica felt confident their relationship would survive their recent rough patches. She relished the closeness so much that she didn't tell Adam when she was ovulating. She usually fended off his advances by saying, "It's not a good time. You have to wait a few days." She knew she could have asked him to wear a condom, but she didn't bring up the subject, and neither did he. She felt pangs of guilt but convinced herself to ignore them. If he really didn't want a baby, she reasoned, wouldn't he monitor her cycle? Most men she knew who didn't want a baby made sure they didn't get their girlfriend pregnant. She decided to interpret his passivity as a sign he was still open to having a baby with her.

When they returned from vacation, Roni noticed the improvement in their relationship. "You seem so much more connected," she said to a radiant Monica. "There's so much love between you. I've seen couples who have been together thirty years who have been working on similar stuff and haven't come as far as you have." Adam squeezed Monica's hand.

For the next few months, they continued to have unprotected sex nearly every day. Neither broached the topic of Monica's ovulation schedule.

Monica felt such a surge of love for Adam that she told Roni in individual therapy she was lucky to have him as a partner, even if they never had children. "So many women look at men and think, 'I love him. He's a good earner. He respects me. We have a good sex life. He'd make a good father,'" she explained. "I've learned that I love *Adam*, not just Adam as a potential father. I'm finally looking at him without the filter of 'Would I be able to have kids with him?'" She hoped such thoughts would soften the reality that they had been unable to become pregnant for a year now.

One evening, encouraged by a productive couples therapy session earlier that day, Monica blurted out the question that was never far from her mind: "What about the wedding date?"

Adam dodged the question and used the opportunity to offer his philosophy on marriage. "It should be a feeling, not an event," he said. "At a certain point in a relationship, it becomes clear that a couple should be married, and a wedding just formalizes that feeling. For now, I'm happy with the way things are." He reminded Monica that he showed his commitment in other ways, such as offering to support her financially if she needed time to look for another job. "I like how our relationship has grown organically. Let's continue on that same path and see where it goes. Do you feel we're ready?"

"Yes, I do," Monica responded, arguing that they had regained the intimacy they had lost by rushing during those early months. "No one is ever 100 percent ready."

In couples therapy a few weeks later, Roni also brought up the subject. She asked them to reflect on how they had improved and what they needed to work on. Monica offered that she felt more connected to Adam and had been making progress in controlling her outbursts when she became angry with him. When Adam admitted that he also felt closer to Monica, Roni asked him when he saw himself getting married. He repeated what he had told Monica earlier: time would tell.

"When do you think would be a reasonable time to know?" Roni prodded.

When Adam shrugged, she asked, "What about the end of the year?"

Adam agreed to talk about it again in December, a couple months away.

Kelly

Dan didn't try to kiss Kelly until their fifth date, but by then she knew she was falling for him. She loved so much about him: his generosity, kindness, strong Christian faith, and the way he cared for his eight-year-old twins, Hunter and Dylan.

Kelly saw Dan every other weekend, when he wasn't with his boys. He usually traveled to Kelly's home in Charlotte, often bringing a bottle of Cambria pinot noir that he had ordered online, because he didn't want her to make the two-hour trip to his small town by herself. Although she hated waiting two weeks between visits, she was pleased that he protected his time with his sons. It would have bothered her if they hadn't been his priority, and she liked that he waited some four months before he arranged for them to meet. She admired his nurturing manner and how he created little father-son traditions, such as playing catch or building bird feeders or benches for the yard. She imagined he would be just as good a father to their children.

Kelly knew within a few months of meeting Dan that she wanted to marry him, and her friends and family remarked that she had never seemed so happy and relaxed. Unlike her first marriage, in which every day was embattled, this relationship didn't feel like work. Dan didn't hide his feelings for her and often dropped comments like "If we get married" into conversation. They also browsed estate rings when they went antiques shopping.

However, if she married Dan, she would have to give up her job and sell her house to move to Brevard, since his sons and job were both there.

Dan hinted that he could provide financially for Kelly so she could focus on her community service projects and eventually care for their baby.

Sometimes the thought of giving up her independence after she had worked so hard to establish her own life scared her. She didn't want a husband who would make her dependent on him or control her finances. After six years on her own, she was confident that she didn't need to be married to be financially stable or emotionally fulfilled. She was proud she could manage her own household and handle chores that men traditionally did, such as changing the windshield wiper blades on her car, cutting the grass, and fixing the lawnmower. She reassured herself that even if she let Dan assume a traditional breadwinner role, she could always work if she wanted to.

Dan proposed in April 2007, almost a year after their first phone conversation. She was still in her pajamas one Sunday morning checking her work email when he knelt beside her computer. "Kelly, you bring out the best in me," he began nervously. "I want to share the rest of my life with you. I want you to be my wife." Then he opened up a jewelry case and with his hands shaking, put on her finger an emerald and diamond estate ring they had seen at an antiques store in Charlotte's South End a few months earlier. "Will you marry me?" he asked.

"Of course I'll marry you," she gushed, jumping up and down.

For the first time in her life, at age forty-two Kelly could freely fantasize about becoming a mother. She thought having a sibling would be fun for the twins, and she also hoped she could play an important role in the boys' lives. She read books about how to be a good stepmom and particularly liked one approach that suggested avoiding the word *step* and thinking of them as *bonus* sons. She felt a genuine connection with the boys and was touched when they asked her to make their favorite homemade macaroni and cheese dinner or argued about who got to sit next to her at restaurants. She delighted in her instant family; a baby would be the icing on the cake.

Kelly increasingly loved the idea of moving to Dan's small town. She imagined her kids would be exposed to travel and big cities, but their day-to-day life would be in a quaint town, where people took pride in historical preservation and still registered for china and sterling at the local shop. She let her mind run wild with decorating ideas for Dan's old Victorian: hanging fruit and pine boughs on the front door, placing big pumpkins on the porch. She thought the alcove down the hall from their bedroom would make a perfect nursery.

Since Dr. Winslow had retrieved nineteen eggs just three years earlier, Kelly assumed she had a good chance of having a baby in her forties. Plus, she ate well and exercised regularly. Dan told her she seemed so healthy that he never could have imagined she'd need medical help to conceive. But she had to educate him about the basic facts of fertility: no matter how much effort she made to take care of herself, there was nothing she could do to make her eggs healthier. In the meantime, she wondered if Dr. Winslow would recommend dipping into her stockpile.

That summer a friend gave her a pair of crocheted booties that came with a little fertility blessing. A couple was supposed to put them between their box springs and mattress and pass them on to another couple after they had conceived. Kelly appreciated the sentiment but left the booties in the hall closet. Given Dan's vasectomy five years earlier, she couldn't try to become pregnant naturally. Although he was willing to undergo a reversal, she read that it could take time for sperm to regenerate. However, she appreciated his confidence that they would find a way to expand their family. "There's no problem that a Petri dish can't solve," he would say. If IVF or Kelly's frozen eggs didn't work, they could try donor eggs or adoption.

Their first stop was the office of Dr. Winslow, who explained that a vasectomy reversal would cost some $6,000, and it would take several months for Dan to begin producing quality sperm. A better and cheaper procedure would be to surgically aspirate sperm through his testicles. They could then inject the sperm directly into her eggs during IVF and freeze the rest of the sample in case they needed more later.

Kelly believed she still had a chance of using her forty-two-year-old fresh eggs and was surprised when Winslow suggested she'd have better odds with her thirty-nine-year-old frozen ones. She thought it seemed wasteful to use up her frozens if she was still producing fresh. Shouldn't she save those for her mid-forties, when she didn't have any chance of producing a baby on her own?

By 2007 Winslow and Yang had produced nearly fifty babies from frozen eggs, and their website now trumpeted, "More babies born from cryopreserved eggs than anywhere else." The vast majority of cases were women who used frozen donor eggs or chose to freeze their extra eggs instead of embryos during IVF. Although the team had frozen the eggs of some hundred women worried about their biological clock, Kelly was one of the first in Winslow's office who had come back to use them.

Her nineteen eggs would be thawed in two batches. About 80 percent would survive thawing, and 66 percent of those would fertilize. She could count on transferring a handful of embryos with each attempt and had a 50 percent chance overall of bringing home a healthy baby. Each transfer cycle would cost $3,500 plus a couple hundred more for estrogen and progesterone drugs. That was on top of the $1,800 fee to retrieve and freeze Dan's sperm. Kelly was relieved that she'd already paid for the bulk of the procedure—the egg harvesting and freezing—nearly three and a half years earlier. She hadn't minded mailing a check for $1,000 every month to pay off the zero-interest credit card. It was a reminder of the smart decision she had made.

After Kelly planned the conception of her child, she got to work on her wedding. Since her family had already scheduled a reunion for July, she offered to organize a dual event in Savannah, Georgia. She found a strapless blush pink taffeta dress and picked out a taupe linen suit for Dan.

On a bright Sunday morning under a canopy draped with orchids on Little Tybee Island, Kelly and Dan pledged their love to one another and then to Hunter and Dylan. "How could I be this lucky?" Kelly thought, choking up as she surveyed the smiling faces surrounding them. She couldn't believe it was possible to feel so much love.

Sarah

Egg freezing had quieted my panic, but it didn't let me forget about babies. That summer, Paul and I took a New England road trip to Tanglewood to see a James Taylor concert, then to Bar Harbor, Maine, and finally to Boston to see his father. When we passed an SUV with the family's bikes strapped to the back, I imagined that was our family. At the concert, I saw our kids running ahead and staking out a place for our big blanket. They would fall asleep while Paul and I cuddled. That evening, I would line up our family's hiking boots by the door of our hotel room. I would place our baby in the arms of Paul's eighty-year-old father.

There's something about long road trips that brings couples closer—and at the same time exposes every raw nerve. My family fantasies were soon punctuated with disagreements over the efficiency of East Coast passing lanes, whether Dunkin' Donuts or Starbucks made better coffee, and what was the point of carefully rolling Paul's shirts in plastic bags to prevent wrinkles if he was just going to throw them in a big pile anyway.

I knew our bickering wasn't about divergent philosophies of driving or packing clothes for a vacation. It was about the future, as in whether he planned to be in it. Within a few days, Paul finally said, "I've noticed there's been an increase in tension." That comment began a two-day discussion on the following facts: I wanted children within a few years. Egg freezing wasn't foolproof. And what was the point of growing closer to someone if ultimately we didn't want the same things? Most people know whether they want kids, and then find the right relationship. We were doing it backward.

Paul said he felt as though it was the right time in our relationship to be talking about all the hard stuff. He *liked* kids, he insisted, and he grew up with good role models. (His parents were married for forty-five years before his mom passed away when he was thirty-five.) He just didn't have any real desire to be a father and liked our lifestyle as it was. He had worked hard for many years and now treasured his time to play more, enjoying romantic lobster dinners and sunset kayak tours in Bar Harbor. "We have so much fun together, just the two of us," he said, rubbing my knee. "If we'd had children with us, it would have been a different vacation. We'd probably be arguing about the kids."

I loved our lifestyle too. We had been dating for about two years, and we had a great time as a twosome. When we traveled to Miami a couple times a year, we swam together in the morning, then set up two tables next to each other and worked side by side—uninterrupted. We soaked in the hot tub after we were done working and enjoyed a glass of wine before dinner. We indulged in long dinners whenever we wanted and read in bed afterward. We held each other tightly during the night. I didn't want to have to let go of Paul to tend to a baby in the other room or be forced to physically lay the baby down between us.

"But what's going to happen in a couple of years, when we're sick of dinners out and vacations by ourselves?" I asked. We'd be wondering what else there was.

We agreed on several things: We needed to figure this stuff out soon because our relationship couldn't survive otherwise. We were both grateful for egg freezing because it gave us time to sort it out. We would start a weekly summit series in which we would read books with different perspectives and talk about the pros and cons of having children.

I was deeply touched that he was so willing to participate in these discussions. Part of me kept expecting him to flee, especially since earlier that summer, when we were in San Diego for my niece's baptism, my mother had shown us the baby blankets she was handing down to me.

"Why are you still here?" I had asked him.

"I love you," he had responded.

Paul *did* love me, and I knew it. He just didn't know about having kids.

We started on a plan to each read for twenty minutes from a different book each week and share what we learned. Our selections were designed to include all perspectives: *I Want a Baby, He Doesn't: How Both Partners Can Make the Right Decision at the Right Time; Do I Want to Be a Mom? A Woman's Guide to the Decision of a Lifetime; The Chosen Lives of Childfree Men;* and *Childfree and Loving It.* The books urged us to think about why we wanted to become parents. To leave a legacy? Pass on the family name? Because we wanted someone to take care of us when we were old? When we thought about babies, did we think of cuddly cooing cherubs or red-faced angry newborns? Did we have any idea how much work parenting would be? Paul protested the tone of *I Want a Baby, He Doesn't* because it assumed the guy was supposed to change his mind. I was rankled by the term *childfree* but conceded it was no less neutral than *childless.*

The "childfree" books boasted of clean quiet homes, leisurely meals, uninterrupted phone calls, close couples, time for personal growth, adult travel destinations, disposable income, early retirement, and a lack of responsibility for the complete safety and happiness of another human being. They promised lives free of stress, hassle, bickering, fatigue, anxiety, and endless boring activities. *Childfree and Loving It* reduced the entire evolutionary imperative to propagate the species to this: "Put starkly, *why* would a person choose the regimented conformity, worry and potential penury that children involve over a life of fun and freedom?" *The Chosen Lives of Childfree Men* warned men not to simply acquiesce to women's desires to have children. "Men Have a Choice" was the title of one chapter. The author claimed that a wife would fall in love with her baby and ignore her husband. If they divorced, he would be forced to sacrifice his financial security to alimony and child support payments.

These "childfree" books seemed to be manifestos, not helpful tools to sort through the decision to become a parent. I couldn't compete with them. The joys of parenting were much harder to describe than the horrors. The rewards were made up of small moments and feelings of pride, love, and satisfaction. It was like trying to explain the feeling of falling in love.

I stressed that it was impossible for Paul to understand women's biological instincts. I had longings that sometimes felt like spasms in my stomach. They were physically painful. The *Childfree and Loving It* book dismissed these as an "acquired emotional attitude" and the products of slick marketing, as if my desire to reproduce stemmed from my wanting the latest expensive stroller.

I felt as if I were participating in a college dorm bull session, hashing out abstract issues that had little application to real life. Of course, dissecting theoretical arguments about having children was no way to decide if you wanted one. This was an emotional endeavor. I wanted a child because I felt an illogical longing for one. It suddenly made sense that Paul wasn't completely sure how he felt. He'd hardly been around children; his previous girlfriends hadn't pushed the issue, he didn't have nieces or nephews, and he had lived alone in Manhattan for most of his adult life. Since he didn't have a deep ache in his gut, I wanted to find a way to reach him on an emotional level. A friend had told me that her boyfriend had changed his mind about having children after falling in love with her niece. Paul needed to see how children could be fun for him. He needed to spend more time with his friends who enjoyed being fathers.

Paul agreed and said my idea was a good next step. I said I would try to organize get-togethers with families we knew. He said he would talk to his male friends who had children and ask them whether they liked being dads.

My niece Kate, who was one and a half, was a good start. When I went home to San Diego a few weeks before Thanksgiving for the funeral of my step-grandfather, I couldn't believe how luscious she was. I crumbled inside when she struggled to say, "Hi, Aunt Sheera." I could actually feel little nurturing neurons firing off when I held her, and my body felt colder when I handed her back to my brother. I loved her best when she was crankiest. She had perfected a look of contempt combined with a dare to make her laugh. It was conditional. It was snarky. It was so me.

I wondered what a combo of Paul and me would look like. Would Claire have his smile and my eyes or thick ankles? I couldn't get over what an incredible concept it was to create another human being with the person you loved most. Whenever I worried about the odds of my frozen eggs working, I often reassured myself that I could still use donor eggs if I couldn't conceive on my own. Sometimes, out of curiosity, I looked up egg donor profiles on the Internet to preview my choices. I didn't find anyone who looked like me, but I saw a lot of pretty women I wished would have been my friends in junior high school. It was a surreal exercise; I could pick a popular girl to be the genetic mother of my baby. I questioned whether it was better to start with a random roll of the dice. My own family gene pool was a stew of heart disease, ovarian cancer, Alzheimer's, obesity, diabetes, depression, and drug and alcohol addictions.

I had always breezily said I was comfortable with "other options." But that was before I had met little Kate. Was there something to genetics? Or did you see any child and look for yourself?

My family joked that they couldn't wait for Paul to see her when he visited for Thanksgiving. "Make sure you come early," my mother instructed. "And make sure she has a good nap," she told my brother. Kate would be our little ambassador for parenthood. And just in case I missed the message, my mother said loudly to Kate, "Well, if your auntie would just get your cousins out of the freezer in New Jersey, you could have a playmate." I corrected her, pointing out that my eggs were in a secure medical storage facility in a suburban Boston office park. But I think she preferred the drama of New Jersey, as if our family's precious genetic material was stashed away in an unlocked warehouse in Newark's South Ward.

When Paul arrived, I was prepared with my baby-exposure plan. I scheduled visits with some of my friends from high school, and Paul watched the dads roughhouse with their toddlers. Another friend told us he was happy to have moved back to San Diego because he loved walking on the beach with his infant nephew.

I remembered what my old straight-talking therapist had said about showing Paul that having kids with me would be a good experience, so when we babysat Kate, I tried to show him how we could enjoy it together and made sure to include him in all the tasks. Paul read her stories and made up a little game in which he gave her a "beep" on the nose. When she cried, he calmed her down by doing a funny dance.

On our way home from dinner, Kate fell asleep in the car, and I showed him how to lift her out of her car seat and carry her to her crib. I watched his face soften as he placed her head on his shoulder and felt the sleeping child against his chest.

I was melting.

"That was fun," he said at the end of the evening. "I can tell your brother really enjoys being a dad."

I didn't ask for elaboration. That was enough for now.

Hannah

In the past, whenever Hannah fell for a guy, she quickly became infatuated. She couldn't stop thinking about him. She wondered when he was going to call. She wondered if he was going to ask her out again. She wondered if he was seeing someone else.

But she had faith that Nate would continue to ask her out, which he did week after week. Compared to other men whose erratic interest made her feel insecure, Nate didn't bring out her anxiety demons. He was consistently affectionate and sweet, and at age thirty-nine Hannah finally experienced what it felt like to be in a stable relationship. For nearly all her adult life, she had thought she was unlovable when, in reality, she just hadn't found someone who was *loving*.

Nate told Hannah she was beautiful. He gave her massages when she had had a hard day. He called regularly. He sent her emails saying he wished they weren't working and could spend the day together. Once he picked her up at work and brought a surprise picnic of wine, cheese, crackers, and oysters. When she was upset about a demanding boss, he wrote in an email, "Just remember there are a lot of people who love you. This is just a job." It made her smile all day.

She didn't have to wonder, "Where is this relationship going?" Three months into their courtship, she knew. They made love for the first time, and Nate asked her to become exclusive.

Hannah also admired how he treated his children, Jason, nine, and Holly, eleven. He coached Jason's baseball team and attended all of Holly's plays. He helped them with their homework, packed their school lunches, and made them chocolate chip pancakes on weekends. Hannah loved to look at the photos placed around his house of him and his kids camping, sledding, and swimming in mountain streams. One of her favorites showed Nate and Jason sitting on a scooter. They looked so exhilarated, and Hannah imagined herself joining in their adventures.

After Hannah's own parents divorced when she was eight, she and her older brother lived with their mom and saw their dad once a week for dinner. It wasn't an unusual arrangement in those days, but it was a sharp contrast from the way divorced dads today are expected to be involved in the day-to-day activities of their children. Nonetheless she was impressed at how Nate had worked hard to create a home for them and shared equal custody with their mother. She later learned that they had lived mostly with him after the separation, before a custody agreement was established. She thought it took a special kind of man to step up the way he did during such a difficult time.

This was the kind of man she could see marrying. This was also the kind of man she could see having a baby with. While some women are in touch with their maternal desires their entire lives, others don't feel them until they approach the end of their fertile years. However, some women's longings are activated when they fall in love. Hannah felt as if a switch had been flipped inside her. Before freezing her eggs, her fertility had seemed like something she was always in the process of losing, and mourning. She never could tell how much she wanted a family and how much she was reacting to fear. Now her gut was clearly saying, "Wow, this is totally possible!"

Hannah was nervous that Holly and Jason would see her as competition for their father's attention, so she insisted on being integrated into their lives very slowly. She never slept over at Nate's while they were there and was careful to respect their time alone with him during his custody weekends. But over time she relaxed and started to enjoy being included in their family. They all sang camp songs in the car and told silly jokes at dinner. Both kids were friendly, and she was especially touched when Holly opened the door for her at restaurants.

Hannah liked the idea of being a stepmom, but she knew the role wasn't a substitute for being a mom. She had already missed so many of Holly's and Jason's early formative years that the best she could hope for was to be a good friend; their primary bond would always be with their biological mother. Besides, Hannah knew well the limitations of stepparenting; her mom had remarried shortly after getting divorced, and Hannah had become close to her stepdad, who often took her fishing and camping. But when that marriage ended seven years later, when Hannah was in her late teens, the man abruptly disappeared from her life. Her mom had since remarried, but the stepfamily tree seemed so fragile.

Four months into their relationship, she and Nate had yet to broach the subject of having a baby. Hannah was terrified to bring it up. She didn't know what she would do if he didn't want any more children, since she was already falling in love with him. She hoped that the longer she waited to talk to him, the harder it would be for him to tell her no. He would be too attached. On the other hand, she too would be more attached, and a disappointing answer would be harder for her to hear.

Finally, over dinner on New Year's Eve, she choked down the lump in her throat and forced herself to tell him about her decision to freeze her eggs. "I still want the chance to have a child," she explained, her shoulders tightening. "I want to try to have a baby when I get married."

"That's good to know," he responded sincerely. "Being a parent is one of life's greatest blessings. If you want it, you should have the opportunity to have it." Hannah felt her stomach drop as she tried to interpret his response. It was cordial but impersonal and mentioned nothing about whether *he* wanted another baby.

Hannah hadn't expected Nate to say whether he was open to having a child with *her*, since it was too early for that conversation. After much rumination, she took comfort in what he *didn't* say. He didn't say, "Well, I'm done with all that." She hadn't received the answer she wanted, but she *had* succeeded in putting the subject on the table and was content to let it sit there for a while.

For someone who had tried to convince herself that she'd be fine if she never had a child, actually owning her desire and communicating it to her boyfriend was a big step. It was also a huge departure from her past approach of going along with what her boyfriends wanted and hoping they didn't dump her. For once, Hannah had a backbone—and a voice.

Monica

Monica had spent more than a year trying and failing to become pregnant. When she suggested to Adam months earlier that she should see a fertility specialist, he had resisted the idea. He told her it would lead to more stress, and that was the last thing they needed.

She tried to manage her anxiety by telling herself they could revisit the subject after they set a wedding date. Still, her family could be counted on to pepper her with questions, such as "Are you pregnant yet? You're almost forty!" Forty seemed like such an ugly number, and it was waiting for her less than a year away.

Monica wondered if she should dip into her supply of frozen eggs, which were nearly three years younger. "Why *not* thaw them?" asked her sister, who had given birth to her two children when she was in her late twenties and early thirties. "That's what they're there for. You don't want to waste another year of trying." In the meantime, her sister reassured her that she still had a chance of conceiving naturally and reminded her of their family friend who had given birth to her first baby when she was forty-one. Monica was confused: Was she supposed to feel comforted or freaked out?

Even though Adam had dismissed her idea of seeing a specialist, Monica decided to visit another ob-gyn. She thought she should be tested, in case the problem was treatable, such as blocked fallopian tubes or trouble ovulating. She also wanted another opinion about her recurring chest pains. "Don't assume it's psychological," Roni had urged her.

Back in therapy, Roni asked Monica to think deeper about her single-minded quest to have a baby. "Is it pressure or a real desire?" the therapist questioned. "Or just a goal?"

"No, I really want a baby," Monica insisted.

The therapist, who knew her client's mind-set well, persisted. "Monica, is this just another thing to tick off your checklist?" she asked. Roni was silent, letting Monica squirm over the question of whether she regarded having children as an important, meaningful experience or thought it was simply expected of her. Wasn't that what high-achieving girls from affluent families did? Monica knew few women who had chosen different paths.

Then Roni took a different tack. "Would you have kids if you were single?" she asked. Monica was so surprised by the question she nearly fell off the couch. She had never seriously thought about that option before, even when her sister half-joked that women should just hit the sperm bank and then worry about finding a partner later.

"No," she responded. "I want to share the experience with someone. I want a family, not just a kid. I want it to be about something we created together."

"Well, it's just something to remember," Roni said. "Just think about what you truly want."

Monica's new ob-gyn performed a pelvic exam and asked a nurse to draw her blood. She also ordered an ultrasound to check for fibroids and another test to see if her tubes were blocked. Then she handed Monica a semen collection cup so she could take it home and ask Adam for a sample to be tested.

"Uh, we're not ready for that," Monica stammered. The doctor simply told her to bring it back when they were.

Her fertility tests all came back negative, and the doctor couldn't explain her chest pains. Monica hated the helpless feeling that there was nothing she could do to help herself get pregnant. The only option left was to test Adam's sperm. But she was hesitant to ask him for a sample and had hidden the semen cup in the back of a nightstand drawer. She planned to present it to him at the end of the year, after their big session with Roni. That's when he had agreed to talk more about his feelings on getting married.

Monica had tried to endure the wait by putting on a brave face and meditating every day to contain her stress so it wouldn't lead to fights. But she couldn't tell what Adam was thinking, and she had worked herself into such a nervous state that she had no idea what her gut was telling her.

She wondered if Adam might be more willing to get married if she downplayed the wedding part and suggested eloping during the Cancun vacation they had planned for January. Later they could throw a small party, and she would wear a pretty cream dress instead of a big formal gown. They were *still* engaged, she said to herself, trying to nurture a seed of hope. She reassured herself that Adam was simply a nontraditionalist who had to do everything on his own time frame. He was process-oriented; she was goal-driven. They just had to find a way to accommodate both perspectives, like an exercise in therapy.

At other times, tears unexpectedly streamed down her cheeks.

During her private counseling session, Monica reminded Roni to ask about the deadline during their next couple's visit. Even though some therapists coach couples on how to talk to each other, Roni was happy to play mediator in this case. She wanted Adam to freely share his feelings about marriage rather than react to Monica.

Finally, the date of their couple's session arrived. "It's the end of the year," Roni began, turning to Adam. "A few months ago, we had discussed a time frame in which you'd be willing to talk about what you wanted to do moving forward. How do you feel about where things are at?"

Monica's body froze as she waited for his answer.

"I'm not ready," he answered. "I don't feel I can give Monica the love she deserves."

Monica sat there, stunned. She had prepared herself for the first part. But his not loving her enough? That had never been an issue before. This conversation was supposed to be about syncing up their timelines.

"Monica, how do you feel?" Roni asked.

"I don't know," Monica responded, barely able to breathe. "I'm not sure what he means."

"Adam, do you want to expand on that?" Roni asked Adam.

Monica stared straight ahead as she listened to the man who had proudly surprised her with a ring sunken in white frosting just two years earlier recycle his tired arguments about why he now didn't believe in marriage.

"Well, we're almost out of time," Roni finally said. "Is there anything else either of you want to share?"

Adam shook his head.

Monica said, "No."

They walked silently to the car.

Later that night, Monica tried to ask Adam again, in case she had misheard his bombshell. "That session was upsetting," she said in her best nonblaming voice, as she had learned to do in therapy. "I heard that you don't feel you could love me in the way I need to be loved." He dodged the question by saying he needed more time to make a decision. Monica couldn't think of anything else to say.

During her family's annual New Year's weekend at the shore house, Monica walked around like a zombie, as she tried to come to terms with the painful truth that Adam wasn't fighting to keep her from leaving. "I don't know if I can give you more time or even if I want to," she told him.

"Maybe I can work through it," he offered.

"Would you wait, if you were me?" she replied.

"Probably not," he replied.

5. *Doubts*

Sarah

Three years after Christy Jones's entry into the egg-freezing business prompted the American Society for Reproductive Medicine (ASRM) to discourage doctors from offering the procedure to Clock Tickers, the professional organization had refused to soften its stance. In fact at the 2007 annual meeting in Washington, D.C., the practice committee chairman, Marc Fritz, announced that the society was reaffirming its position. "Our concern is that women may regard freezing their eggs as insuring their fertility," he told a packed press conference. "Existing medical evidence doesn't justify that conclusion."

His voice grew hoarse as he tried to make his point to the young British reporter who had asked about the success rates of frozen eggs. "In what center? In which doctor's hands?" Fritz fumed, enunciating carefully. "Patients can't make informed decisions without accurate information!"

I raised my hand. "What about all the improvements we've been hearing about?"

Fritz, a fertility doctor at the University of North Carolina Women's Hospital, admitted that although the technology was better, the promising studies came from a handful of pioneering centers. Many doctors who offered egg freezing had yet to even thaw eggs and make a baby, he said, and they often quoted those glowing success rates rather than their own records. Plus, most of the data that were available were based on the frozen eggs of younger women, not those of women at the end of their baby-making years who were the likely patients.

The data weren't consistently reassuring. I had spent the past couple of days at the conference listening to presentations of egg-freezing studies. One team achieved seventeen pregnancies from 155 eggs and estimated it took nine frozen eggs to make a baby. Other researchers said they averaged ten eggs per birth, and another group claimed they needed about twenty eggs for each baby.

Making the picture even more complicated were the different designs of each study, which significantly shaped the results. For example, researchers who offered fertile women free cycles in exchange for being part of a study saw higher rates than those using the eggs of infertile women on their third IVF treatment. Those who tried the newer flash-freezing method of vitrification seemed to have more success than those who used slow freezing, but it was hard to make any conclusions, given all the other caveats. Finally, a lot depended on where you practiced medicine. In Italy, where embryo freezing was forbidden, doctors were allowed to thaw and fertilize only three eggs at a time to avoid creating multiple pregnancies. (Similar restrictions existed in Germany.) However, in countries where embryo freezing was allowed, researchers could freeze and thaw a patient's entire harvest and pick the best embryos to transfer back to her uterus.

I understood that the ASRM couldn't wholeheartedly endorse a procedure with such uncertain success rates. But it wasn't as if other fertility advances, such as IVF, ICSI, embryo freezing, and embryo genetic screening, were good bets from the beginning. Yet the techniques weren't saddled with the designation "experimental," and the medical establishment certainly didn't dictate to doctors whom they could treat.

Had egg freezing been unfairly scrutinized? Frankly, it's hard to say because egg freezing *is* different from any other fertility treatment. For one, many doctors don't like offering it. The concept of performing a surgical intervention on healthy women as a hedge against future infertility runs counter to doctors' "First do no harm" mandate. (Women who show signs of early menopause might be seen as exceptions, but certainly not thirty-six-year-old women whose only "illness" is that they will someday get old.) Then there is the psychological burden. No matter how strongly physicians caution patients that egg freezing might not work, they are never sure patients really believe them. As Michael Tucker told the Associated Press in 2004, "People don't hear what you say. They hear what they want to hear and wind up paying for a service which they believe is much closer to certainty than it really is." Doctors know that patients who wind up empty-handed can't help but blame them on some level. Frankly, they would rather be heroes to infertile couples who show up with nothing to lose in the first place. That's the same reasoning behind why many doctors will offer egg freezing to cancer patients but not Clock Tickers. It takes away the pressure to stand behind the science when patients have no other options.

Freezing is also different from other treatments because doctors can't quickly gain experience but often have to wait years for patients to return to thaw their eggs. Since many clinics don't have track records, patients are betting that their doctors will have more practice turning eggs into babies by the time they're ready to use them. Given these dynamics, doctors are hardly accountable for the outcomes, and patients are supposed to approach the whole process with low expectations. However, from a patient's perspective, this "Oh well! We tried our best" attitude is hard to stomach when the financial cost and emotional stakes are so high.

Finally, egg-freezing patients have little information to gauge their likelihood of success. When you undergo IVF, you receive real-time feedback on your fertility health: it worked, or it didn't. Even when you freeze embryos, you know whether your eggs will fertilize and flourish. But you have no idea how good your frozen eggs are until you try to use them. Although some doctors are experimenting with new techniques that test eggs' polar bodies to evaluate their chromosomal structures, most physicians can offer only general comments on egg shape, size, and pearliness. If you've never been nor tried to become pregnant, it's hard to know whether you can rest easy by banking twelve eggs or need to go for the gusto and freeze as many as possible to have a good chance of bringing home a baby.

If the reproductive technology field wasn't rattled enough by the very idea of egg freezing, the way entrepreneurs like Jones promoted it exasperated doctors. First, it should be said that doctors in general feel awkward selling themselves, since they believe that practitioners of sound medicine shouldn't have to advertise their services. So imagine their surprise when Jones showed up with her marketing muscle.

After starting Extend Fertility in 2004, she hired a public relations firm, created an educational website where women could take quizzes about fertility facts, and tried to initiate viral marketing by sending email questionnaires about women's fertility to friends and acquaintances and asking them to forward these to other interested women. She also organized direct mailings and email blasts to potential clients. In one instance, she bought a mailing list of female subscribers to *New York* magazine and sent them postcards saying "Set your own biological clock." (Subsequent cards urged them to "freeze with a friend.") She and her staff tried to come up with other innovative strategies, such as contacting divorce attorneys and psychologists, posting banner ads targeting women of a certain age on the Internet dating site Yahoo! Personals, and handing out cards and egg-shaped soaps at Jewish dating events called Matzo Balls. They even sent promotional materials to aging Hollywood stars, including Salma Hayek and Jennifer Aniston, in the hopes that they might become future poster girls. (Rumors swirled that Aniston froze in 2008, but they were never confirmed.) They also sent postcards directly to the people women talked to about fertility—the ob-gyns— that explained the benefits of egg freezing and suggested they share the information with their patients.

In some ways, criticism of Extend Fertility—and Jones—was unfair. Several doctors dismissed her as a graduate student running amok with a business school project rather than acknowledging her as the seasoned entrepreneur she was. Also, she became partners with reputable clinics and coordinated training of their staffs who ultimately were responsible for the integrity of the science and well-being of patients. And even though the word *commercialization* has a distasteful ring to it, many voices have argued over the years that health care could benefit from being treated as any other good and service on the open market. The selling of a medical procedure wasn't necessarily a bad thing. By stirring the pot, Jones forced the ASRM to focus on the issue, and the resulting publicity prompted doctors to defend their standards (and hopefully even raise them) and encouraged women to be more discerning consumers.

However, Jones did make it appear that egg freezing was further along than it really was. Her ad slogan "Set Your Biological Clock" was snappy and creative, but it was misleading. In the best of cases, with experienced doctors and a lot of money and cycles, a woman might be able to *slow down* her reproductive clock by a couple of years, until she had to think about her life span and aging clock. Similarly, in the spring of 2004, Jones could be heard on morning talk shows making what many critics considered overreaching claims about the potential of the procedure: that women could expect a pregnancy rate approaching their natural peak fertility rate of 30 percent. She was referring to the Italian success rates, but she left out an important detail: at the time none of her clinic partners had thawed any eggs and made a baby.

Even when she funded research with Reproductive Medicine Associates of New York that produced five frozen-egg babies, the research was subject to criticism. I had been inspired by that study when I heard about it at the Extend Fertility information session in the summer of 2006. To me, it was proof that egg freezing was for real, and I often reread the press release, which boasted of a thawing survival rate of 86 percent, a 90 percent fertilization rate, and a 26 percent implantation success rate. Yet only when another doctor (who clearly wasn't an Extend fan) prompted me to look more closely at the actual study did I realize what I had missed. The eggs that were frozen were those of donors with an average age of twenty-eight, which were then thawed and transferred to infertile patients in their forties. The doctors transferred twenty-three embryos to four women (that's an unusually high number of six embryos at a time) to achieve those pregnancy rates. When I first read about the study, I imagined that I could easily be one of the three out of four women who got pregnant. But now I realized I was way out of my league: I wasn't in my late twenties, and I didn't have twenty eggs to play with. So far, I had six. Based on the laws of attrition, I might have a shot with one or two embryos that eventually would be transferred to my womb, not the five or six at a time that accounted for the success rates.

Extend and RMA deserve credit for creating babies in a sea of inexperienced shingle-hangers and showing that the science worked. (Over the next couple of years Extend would add nine more babies from five more clinics.) The problem was that they didn't quite show how it worked for someone like me, or my peers sweating over our Chardonnays during well-organized information sessions in fancy Manhattan office buildings.

I bristled at the ASRM's edict that egg freezing "should not be offered or marketed as a means to defer reproductive aging." The underlying message was that women, *especially* women consumed with baby panic, were too emotionally vulnerable to clearly evaluate the risks. The ASRM's presumption that women needed protection from themselves had an unsavory paternalistic feel. At what point did efforts to protect us from making potentially bad decisions interfere with our right to protect ourselves with potentially good options?

However, the society's stance did beg a legitimate question: Did women need to be protected from the likes of inexperienced doctors and overexuberant promoters? Since no governmental agency directly regulates fertility medicine, did it make sense for a doctors' organization to take on the task? Despite its bark, the ASRM actually had no power to determine who offered or marketed what services to which patients. So the organization took a persuasive tack and issued an opinion, titled "Essential Elements of Informed Consent for Elective Oocyte Cryopreservation," that urged doctors to go over thirteen points with prospective patients before beginning treatment and indicate that they had been counseled in their medical records. The expectation that providers of egg freezing would tell their customers every possible reason they shouldn't buy their service was counterintuitive, in the same way cigarette manufacturers have to pay for public health announcements to dissuade people from smoking. But for the most part, the points made sense: women should know about possible side effects of hormone medications and the surgical risks of egg retrieval; they should know that they have to pay for storage; they should know that many studies were done with younger women and may not reflect realistic odds for those over thirty-five; they should know there is a possibility that none of their eggs might survive thawing or that they could be lost in the lab; they should know what might happen to their eggs if they die. And the kicker: they should know the "potential risks of basing important life decisions and expectations on a limited number of cryopreserved oocytes."

So the question remained: Could the ASRM's "Essential Elements of Informed Consent for Elective Oocyte Cryopreservation" referee reason and emotion and help us make better decisions?

The document was a laudable effort, but the doctors' organization could also be accused of cherry-picking the most discouraging statistics to make *its* point. For example, it said that doctors should quote actual success rates from their own centers and not refer to "likely" or "possible" success rates achieved from other researchers' studies. In the absence of their own statistics, they should quote a live-birth rate of 2 percent per thawed egg that was cryopreserved through slow freezing, a rate taken from a review of all reports of frozen egg births since 1997 that was published in the ASRM's academic journal *Fertility and Sterility* in 2006. There are two problems with this figure, however. First, using a per-egg analysis is an uncommon method of estimating success rates because it makes those rates seem alarmingly low. Since doctors work with a batch of eggs at a time and fertilize and transfer several embryos to a woman's uterus, they usually quote a success rate per *total* embryo transfer. For example, the meta-analysis did a comparison with standard IVF cycles using fresh eggs and found a per-egg success rate of only 6.6 percent but a total embryo transfer rate of 60 percent. Second, the ASRM opinion quotes a figure that reflects all egg-freezing births since 1997 and does not acknowledge that the rates jumped significantly when slow-freezing methods improved in 2001. It also includes success rates using eggs that were not injected via ICSI, which is standard procedure today. Live-birth rates per *injected* eggs after slow freezing are 3.4 percent per egg and 21 percent per total embryo transfer. As for the eggs frozen between 2002 and 2004 (which reflect the huge improvements to the slow freezing formula), the rate jumps to 5 percent per egg and 32 percent per embryo transfer.

This opinion (reissued in 2008 with the same information) also doesn't represent how most eggs are frozen today. While many clinics still use slow-freezing methods, many have switched to the more successful method of vitrification, which the opinion estimates has a 4 percent success rate. In another article published around the same time in *Fertility and Sterility*, the ASRM's practice committee even lauded vitrification's potential: "In humans, post-thaw survival rates of vitrified oocytes have improved and fertilization rates are beginning to rival those of fresh oocytes."

Even the ASRM seems unsure whether to be cautious or hopeful. Yet their organization's opinion matters. The media may have failed to reflect such warnings in their chirpy coverage of egg freezing, but other doctors took note. The American Congress of Obstetricians and Gynecologists followed the ASRM's lead in 2008 and issued a similarly bearish statement about the potential of egg freezing. The problem, though, is that when ob-gyns are asked about egg freezing by patients during their annual exams, they don't have the time or the information to give nuanced answers about the "state of egg freezing today." Anecdotal evidence suggests that many quickly dismiss it as an option, saying simply, "It doesn't work." Patients leave with an inaccurate view of a procedure that might benefit them. They also leave without referrals for the good doctors out there.

The other problem is that the ASRM dissuades women younger than thirty-five from freezing, arguing that there is a "relatively high likelihood" that they won't need to use their eggs because the "large majority of women marry by 35 and have a low incidence of childlessness." That opinion, however, ignores the reality that younger women who are interested in freezing are the ones who don't foresee starting a family by age thirty-five. And even if they slide in under the deadline, who's to say they will get pregnant right away or won't have trouble in their late thirties trying to conceive a second child? After my breakup at thirty-two, no matter how quickly I reentered the dating market, I knew that the odds of landing a good man, getting engaged, married, and then pregnant before my fertility started going downhill weren't great. In some ways, *that* was a riskier gamble than betting on my eggs working later. It would seem prudent for the fertility field to encourage younger women to freeze when they have the best possible odds of success rather than shake its collective head at the post–thirty-five crowd for waiting until the last minute *after* it becomes clear their men aren't coming to the party.

It was fascinating to see how the controversy played out overseas. To truly understand the story of egg freezing, I visited its birthplace in Bologna, Italy. That was where the team of Eleonora Porcu and Raffaella Fabbri created the first frozen-egg baby using ICSI in 1997 and has since been responsible for hundreds more. They developed egg freezing because the Vatican and much of the Italian public weren't comfortable with the idea of freezing embryos. The Italian government debated the issue for decades before finally outlawing the practice in 2004. Fertility doctors needed to find an alternative, or else IVF patients would be forced to waste all their eggs that weren't immediately fertilized.

Bologna was also the site of the second World Congress on Egg Freezing, which I attended in December 2007. I listened to two days of presentations on everything from whether you could freeze embryos made from frozen eggs (a so-called double freeze) to whether the miscarriage rate was higher for frozen-egg pregnancies and how to set up a global registry to monitor the long-term health of babies born from frozen eggs.

But the highlight was a presentation clumsily titled "Egg Storage for Deferring Reproduction: Medicalization of a Social Issue" by Dr. Nicole Noyes of New York University. "Who are we to judge a person's decision to delay childbirth?" she questioned, pointing out that society embraced a double standard that tolerated older fathers more than older mothers. Women should have the same opportunity to have children when they are ready, she said, referring to her own decision to delay starting her family until her mid-thirties so she could complete her requirements to become a doctor.

The room pounced. One participant asked Noyes if American women knew when their fertility ended. Many do, she said; however, she still saw patients who overestimated how much time they had left: "They say, 'I'm forty and fit.'" Another doctor asked her how old was too old to freeze. Noyes recommended that patients freeze their eggs before their thirty-ninth birthday and use them by their early forties. That way, if their frozen eggs don't work, they can still try IVF with their fresh eggs.

I was intrigued by Noyes's "mitigated risk" approach. The rub, though, was that it didn't really buy you *that* much time. I shuddered to imagine how much each year of "freedom" cost, if you divided what you paid for the procedure by the number of years until you gave birth. Of course, it's impossible to quantify the psychological worth of knowing you had eggs put away, whether you ended up using them or not.

Noyes's recommendation shed light on the strange reality of egg freezing. You were supposed to plan your life as if you wouldn't need your frozen eggs, but when you had eggs stashed away in a freezer, that goal had the same flimsy intentionality as starting a diet and keeping your fat jeans in your closet. Women froze their eggs because they knew there was a high degree of probability they would have to use them. I'm not sure women were "basing important life decisions and expectations on a limited number of cryopreserved oocytes." Rather they were fully realistic that they may have to *resort* to those oocytes because that's how their life might turn out.

I stayed in Bologna after the conference to meet with the Italian fertility doctors who had the most experience with egg freezing. I was curious to learn whether the technology had changed Italian views on when to have children, especially in a country with 1.3 children per adult couple, one of the lowest birth rates in the Western world. I had read that many Italian women start their career late or struggle to afford child care and end up having only one child. Surely Italian women were freezing their eggs by the dozens.

Dr. Andrea Borini, who ran a private clinic and was responsible for more than a hundred egg-freezing births, saw patients from New York and Israel, yet he hadn't frozen the eggs of even one Italian Clock Ticker. Italian doctors froze mostly for IVF and the occasional cancer patient. Dr. Borini blamed the lack of interest on what he called a "Latin mentality," which correlated libido with fertility. "If they still have a high sex drive," he explained, "they think they can have babies."

I was speechless. Did Italian women not feel the baby panic, or had they not spent much of their adult lives being hit over the head with scary fertility figures? They didn't even have the variety of options that American women had, since their government had outlawed donor eggs and sperm. Their own IVF clinics were crowded with women in their forties. How could they not think about the biological clock? Did they even know of the medical treasure they had in their own backyard?

After days of camping out in the dim hallways of the University of Bologna public fertility clinic, I finally met with Dr. Eleonora Porcu, a harried woman with a frizzy bob who seemed to be always carrying vending-machine espressos. She too had been contacted by foreign women seeking egg freezing, but she had turned them all down. "To deliberately undertake treatment with surgical intervention and to pay a huge amount of money for hypothetical future use is not appropriate," she said emphatically. "It's really nonsense to freeze after forty, even after thirty-five or thirty-six. The intrinsic quality of the eggs after a certain age is inadequate, and you expose eggs to additional stress."

But her objections were mostly philosophical: she didn't want to promote an alternative that encouraged women to postpone having children. She said she thought it was unfair that women felt they had to give up their most fertile years to go to school and start their career, especially when their baby-making window was so brief. A woman who found a partner shortly after college and started her family at the relatively young age of twenty-five had only about a decade to complete her family. Egg freezing might seem to be an advance for women, Dr. Porcu acknowledged, but it was really a defeat in the long run. Increasing the need for surgery was not progress. "Now we're becoming prisoners of assisted reproductive technology," she said. Instead we should fight for more protection for mothers in the workplace.

Porcu's call to revolution was admirable, but what were the rest of us supposed to do in the meantime? I wonder if she sensed my anxiety because she suddenly softened her tone. "Now, I don't want to sharply cut this opportunity. It might be an additional tool for women to be the owners of their reproduction," she said. But she felt that it was too early to commercialize the option, stressing that women must beware of the "merchants of reproduction" whose goal was to make money.

Would she have felt differently if motherhood hadn't worked out for her? She loved to take long drags on her cigarette as she told her own story, how she was married at twenty-five but had to work such long hours that she didn't have her first and only child until she was thirty-five, after her husband threatened to leave her if they didn't start a family. She had successfully hidden her growing belly from her colleagues and kept working even after she had gone into labor so she wouldn't be accused of being unreliable. "In this type of society, dominated by men, you have to double your efforts," she said. She gave birth to a baby girl in August, when many of her coworkers were on vacation. When they returned, many were shocked to learn she was even pregnant.

Porcu was certain that if she had had a baby at age twenty-five, she wouldn't be where she is today. Then her face relaxed, and she said that having a baby was the most wonderful experience of her life. She reflected wistfully that if she could live her life over, she would have started earlier and had three children. "It's less difficult than people imagine," she said. "Young women should have more courage, even if they don't have the money or job." She had wanted to try to have a second baby, but she was so busy at work she kept putting it off. "Then it was hard to get the momentum back," she said. "The moment had passed."

6. *Time*

Monica

On a chilly February morning after Adam left for work, Monica moved her belongings out of his yellow bungalow. She tucked the wedding rings they had ordered a year earlier into his nightstand drawer and threw away the semen collection cup her doctor had given her for getting a test sample from him. She was surprised she didn't cry when she carried the last box past the two big flowerpots on the porch to her car.

When the raw grief emerged a few days later, Monica tried to comfort herself by creating a tidy narrative: She had given the relationship her best effort and could walk away without regret. She had enjoyed her years with Adam and was proud of the skills she had learned in therapy. She would learn from her mistakes: she wouldn't move in with someone until she was engaged; she would approach relationships more slowly. Although Adam's declaration that he didn't love her enough stung her to her core, she felt an enormous relief to stop working so hard on something that wasn't working.

A few months later, she decided to begin dating again and reactivated her account on Yahoo! Personals. A man she had been corresponding with before she met Adam contacted her and took her on a couple of dates. Monica liked Tony but felt weird kissing him. She and Adam had shared such a powerful physical connection that she found it strange having to get to know someone else's body again. Tony's stories weren't as entertaining as Adam's. His sense of humor wasn't as sharp. Getting to know him just made her miss Adam more.

Meanwhile her fortieth birthday awaited. Monica told herself it was just another birthday, but she couldn't ignore the significance of the milestone. It was a marker that her life was half over, a measure of what she had achieved and a reminder of what she hadn't. She thought it was strange that she had felt so differently about her thirtieth birthday a short decade ago. She had just finished her MBA, but she hadn't known what job she would take. She hadn't known whom she would marry and when she would have children. She had assumed it was only a matter of time before the details were revealed.

As people take longer than in previous generations to finish their education, find a marriage partner, and become financially stable, there is increased social acceptance of an expanded period of so-called young adulthood. But if it's true that "forty is the new thirty," there doesn't seem to be much tolerance for pushing it any further. Simply put, social representations of *still* unestablished, *still* unattached forty-year-olds become markedly less kind once they hit the milestone. Forty is a clear deadline not only in terms of fertility but also, it seems, in life.

Monica at least took comfort in the fact she didn't *look* forty and had always been tickled when people feigned shock at her age, exclaiming, "You look so young!" But on some mornings, she couldn't help noticing heavy bags under her eyes that made her look tired. She also had fallen arches and persistent sciatica.

The prospect of turning forty had loomed for so many years that by the time the day arrived, it had lost some of its punch. Monica celebrated with her cousins and girlfriends at an Asian fusion restaurant in New York. At 5 a.m., she found herself in the Lower East Side at a karaoke bar screaming into the microphone the anthem of newly single women everywhere: "I Will Survive." By the end of the song, she was laughing so hard that she didn't have any energy left to feel sorry for herself.

Instead she embraced the time-honored tradition of women turning forty: self-improvement. She started meditating every morning, getting acupuncture treatments once a week, attending workshops to learn relaxation techniques, and running three miles a day. Within weeks she was confident the missing pieces would fall into place shortly. "You're going to make things happen," she told herself. "You're going to have a big shift." Her biggest triumph: she didn't ache as much for Adam.

She even felt unexpectedly calm about what forty meant in terms of her fertility. For the past decade or so, that age had been drummed into her head as the endpoint. Now that her natural fertility was presumably gone, she could *really* relax. Whether or not she would have a biological child was completely out of her control. It was too late to freeze any more eggs, and there was nothing she could do to improve her chances with her existing frozen eggs.

Monica knew she would have a baby someday, somehow. But for now, she didn't feel devastated that she likely wouldn't be having one the old-fashioned way.

She also wondered whether her longing for a baby would end once her fertility did. Did your body stop sending you reminder pangs when you were supposedly incapable of producing a baby? Was this God's grace at work to soften the disappointment? Monica couldn't tell because she felt fewer longings for anything or anyone. She didn't feel the intense hunger for a man pressed up against her at night, let alone a baby in her arms. She just wasn't as lonely.

Monica's incoming emails had tapered off at Yahoo! Personals, so she canceled her account and joined Match.com to tap into a new pool of prospects. Regarding the question "Want kids?" Match asked members to choose "Definitely," "Someday," "Not Sure," "Probably Not," or the unambiguous "No" or "No, but yours are okay." In the past, Monica would have selected "Definitely" so she would be transparent about her goals of starting a family and could weed out candidates who weren't serious about settling down. But at age forty, she feared she was suddenly in a romantic "no-man's zone": men who definitely wanted kids would seek someone younger. (Many in that category listed their age limit as thirty-five.) Or they would falsely assume she wanted to have a baby as quickly as possible. On the other hand, she was in the right market for men her age or older who already had kids, but she worried they wouldn't want to start over with a second family. Should she take a chance on a "not sure" man who might warm up to the idea?

Monica selected "Someday" so she didn't come across as baby-crazed. It was partially accurate since she *wasn't* as baby-crazed as before. She worried about attracting men who selected "Definitely," since she thought they would want two children, and she didn't know if she had two babies in her supply of nineteen frozen eggs. She also couldn't assume others were as open as she was to donor eggs or adoption. She had entered a strange new dating reality: she had to find someone who wanted children but who was willing to take a chance on her frozen fertility.

First, though, she had to *tell* them about her frozen eggs. She hated the talk. If she mentioned the topic too soon, she seemed too focused on having kids when she was trying to send the opposite message. When she had broached the topic on previous dates, she often felt like a romance-killer as she gave men a biology lesson about how the procedure worked. She was relieved when they seemed enthusiastic. She didn't know if they were impressed with her foresight, relieved to hear she still had options, or happy that it bought them time.

She liked to joke that if she waited long enough, Match.com would include the question "Frozen eggs?" on all the profiles of women over thirty-five. The idea actually isn't so absurd, considering that Match.com used to ask members, "Willing to adopt?" The site has since eliminated that question and now it asks members only whether they want children, whether they have children, and whether they live with children. But it would be in the interests of online dating sites to acknowledge that older women have many options in baby-making and perhaps include a question along the lines of "Willing to explore reproductive technologies?" or "Willing to use donor eggs?" Many men, especially those who place a high priority on passing down their own genes, simply don't know that women over forty can have children using frozen or donor eggs and *their* sperm.

Monica briefly wondered if she should lie about her age on her profile, since she believed that forty was not a real representation of her fertility. Should she say she was thirty-seven, which was the age of her frozen eggs? But she didn't want to start a new relationship with a lie, so she typed in her real age.

Just as she had anticipated, her dating windfall *was* different from the first time she had tried Internet dating nearly three years earlier. Although she was happy to receive a dozen or so daily electronic "winks" or emails during the first week, they were mostly from men in their mid- to late forties and early fifties who were evenly divided between wanting children and being "not sure." She had hoped to hear from more men in their mid-thirties, who she imagined would be more motivated to start families. Sometimes she hated the way online dating zoomed in on a person's age. She didn't think younger men would be thinking about it so much if she met them in real life. They would be thinking more about how to get laid.

Monica was frankly wary of older "never married" men who said they wanted children. If that was the case, why hadn't they started their family already? She stopped that train of thought, though, when she realized they could have the same concerns about her.

Within a few weeks, Monica had a decent mix of Match suitors and was going on several dates a week. Some men asked for second dates, some didn't follow up at all, and others grew intimate quickly, professed their love, and never called again. Monica tried to take it all in stride. At one point, she was communicating with six prospects and wondered if she'd have to pull out her old spreadsheet. But she quickly dismissed the idea. That was not something the New Monica would do.

Kelly

Before Dr. Winslow would thaw Kelly's frozen eggs, she had to undergo nearly as much prep work to have her eggs put back in as she had originally withstood to have them taken out. First, there was a vaginal ultrasound to make sure she had no polyps, cysts, or fibroids on her ovaries. Then he injected dye into the lining of her uterus and viewed the image on a monitor to see if she had any problems there. Finally, he performed a "trial transfer," in which he used a catheter to practice depositing an imaginary embryo into her uterus. When Kelly's next period started, she began wearing estrogen patches to artificially build up her endometrial lining so the embryos would have a place to implant. After two weeks, she added progesterone shots for four days to maintain her lining's thickness until transfer day, which would have been the time her body normally ovulated. But instead of receiving sperm during sex, her body got premade embryos.

Kelly was flush with anticipation as she and Dan packed their bags for the seven-hour trip to Jacksonville. Despite Dr. Winslow's 50 percent estimate of her odds of success, she was sure that within the next few weeks she'd be pregnant. She considered nineteen eggs a treasure trove of possibility. The whole process seemed so easy: Meet man. Get married. Thaw eggs. Give birth. The old-fashioned way of having babies, on the other hand, seemed so mysterious.

Friends had offered to let them use a vacation condo on nearby Amelia Island, which was the perfect place to recuperate. As soon as they arrived in Florida, Dan had to undergo outpatient surgery to aspirate his sperm. Then Dr. Yang thawed nine of Kelly's eggs, injected them with Dan's sperm, and froze the rest of the semen sample for subsequent tries.

The next twenty-four hours were torture as Kelly worried and waited for the results. Until this point, she had been banking on the promise of her eggs. She didn't know if they would survive the freezing. She didn't know if Dan's sperm was viable. She didn't know if embryos would develop.

So Kelly shrieked with delight when Dr. Winslow called to tell her that three of the nine eggs had developed into quality embryos. At first, she was disappointed with the dramatic attrition, but she felt better when she learned the survivors were Grade 2. Embryos are ranked on a scale of 1 to 4; Grade 1 embryos are perfect-looking and have the best chance of implanting, while Grade 4 embryos appear dark and rarely lead to pregnancies. Grade 2 embryos have slight imperfections but are considered high quality and accounted for 90 percent of babies born through IVF at Dr. Winslow's clinic. That night Kelly felt a deep contentment as she imagined the three little embryos she and Dan had made dividing and multiplying in the incubator in Dr. Yang's lab.

The next day, she signed several forms acknowledging that she was participating in a research study. She also indicated that she understood Dr. Winslow was about to transfer three embryos. In a younger woman, it was risky to transfer three embryos because the patient might become pregnant with triplets. But it was unlikely that all three of those made from Kelly's thirty-nine-year-old eggs would successfully implant. Winslow estimated her chances of twins at less than 10 percent and her chances of triplets at less than 5 percent.

Kelly had signed another form asking if she and Dan would accept "selective reduction" if she became pregnant with multiples. This is medical parlance for terminating one or more fetuses if they suffer from birth defects or are crowding out the others in the womb. Winslow used the answer to determine how many embryos to transfer. If a patient pushed him to implant more embryos than he recommended, he preferred the patient be open to selective reduction. If she wasn't, he would be more conservative. The answer wasn't binding, but it revealed the mind-set of his patients. Kelly had a visceral reaction to the question. She had waited so long for children that she didn't think she could abort any of the fetuses and checked "No" on the form.

The embryos were two days old when they were transferred back into Kelly's body. Winslow placed a wedge under her bottom to tilt her pelvis upward, inserted a speculum into her vagina, guided the catheter through her cervix, and squirted the fertilized eggs onto the top of her uterus. Kelly could see tiny black and white spots on the nearby monitor. She suddenly felt a heavy responsibility for the three life forms in her body. Winslow asked her to stay on the transfer bed for a half-hour, and Kelly wondered what would happen if she stood up quickly or went to the bathroom. Had the embryos already burrowed inside her, or were they still floating around? "Don't worry. They're not coming out, even if you cough or sneeze," said Winslow, explaining that the embryos were sandwiched between two surfaces and forming sticky attachments to her uterine lining.

As she lay still, she felt her body relax. She had sought the help of science and hoped God would take over from there. All she had to do now was be an incubator and follow Winslow's instructions to stay in bed for the next day except to eat or go to the bathroom. She couldn't have sex for at least three days to avoid any uterine contractions, which could theoretically expel the embryo. She also wasn't supposed to exercise. Even a brisk walk could raise her body temperature and mess with her hormones.

Kelly wouldn't know if she was pregnant for two weeks; it takes that long for an embryo to implant and produce enough of the pregnancy hormone to be detected in a blood test. Part of her enjoyed having medical permission to take it easy, but since she was an avid runner, she was frustrated at having to sit still for so long; being told she couldn't run made her want to do it more. When friends asked why they hadn't seen her at the gym, she made up a story that she had hurt her knee and needed to let it heal. When others asked why she had been gone a week, she fibbed that she had gone to Charlotte to get her house ready to put on the market.

Kelly wanted to tell her friends she might be pregnant, but she and Dan had decided to share the news of the transfer with only a few family members. They didn't want her fertility adventures to be the talk of their small town. They also didn't want the news to travel back to the twins, in case the results were negative.

She tried to stay calm, but her mind wandered, and she began to grow increasingly worried about whether she had done everything she could to give herself the best chance of success. Once she forgot she wasn't supposed to put any strain on her lower abdomen and picked up two heavy bags at the grocery store.

Her biggest source of anxiety was a strange sensation she experienced in her abdomen for the first four days after the transfer, as if someone were grabbing the lining of her uterus with a pair of pliers. "Is it a pinching feeling?" Winslow's nurse asked. "That's actually a good sign." Kelly wondered if she was feeling the embryos burrowing in, like sand crabs at the beach. But she panicked when she noticed tenderness under her arm and felt a canker sore developing in her mouth, symptoms that usually meant her period was around the corner. One friend she had told about the transfer tried to reassure her that early signs of pregnancy were similar to PMS and could easily be confused. The nurse also told her not to worry. But how could she not?

She tried to distract herself from obsessing over every physical sensation by staying busy and focusing on house projects, her volunteer work, or the boys' school activities. Occasionally she experienced a few blissful moments when she actually forgot she could be pregnant. They didn't last long, though. She and Dan ate most of their meals together and usually prayed beforehand. The blessings always contained requests for a healthy pregnancy and for God to nurture their little embryos. In her nightly prayers she pleaded more boldly for help.

Kelly had heard of women taking home pregnancy tests before undergoing the official one at their doctor's office to see if it detected the faint pregnancy hormone, but she refused to do this. There was a chance it would be negative by mistake, leading some women to stop taking their medications too early. But the main reason Kelly avoided using a drug-store test was the same reason she had avoided her fertility tests prior to egg freezing: she didn't want to know bad news any sooner than she had to.

Hannah

Hannah had always heard that the older you were, the less time it took to decide if you wanted to marry someone. You were supposed to know yourself so well that it quickly became clear if you'd found your match. Hannah thought she and Nate were compatible, but despite her desire to be a mother, she was in no rush to hurry their relationship.

Frankly, she wasn't ready to give up being single. Hannah had longed for a boyfriend for so much of her life that she had never realized how attached she had become to being unattached. She needed time to get used to the idea of merging lives with another person as a girlfriend, let alone a wife. Some nights after work, she was delighted to come home to her quiet Craftsman house that she had painstakingly decorated, painted sage green, and kept very orderly. She would fling off her shoes, eat dinner on the couch, and watch whatever she wanted on TV. On weekends when Nate had his kids, she relished her freedom to see a movie with a girlfriend or invite a friend out for drinks on the spur of the moment.

Of course, she could always do those things, with or without a boyfriend, but she wasn't used to coordinating her schedule with another person. She liked to fill up her days well in advance, and Nate wasn't a planner. If she overbooked herself, she risked not seeing him. If she kept her calendar open, she might miss the chance to set up something with someone else. Sometimes she felt obligated to invite him to group gatherings she didn't think he'd enjoy and wished she didn't have to. On the other hand, she sometimes felt hurt when he didn't include her in outings with his kids.

They were working out the normal kinks of being a couple. But on occasion, she reflected how simple her single days had been, when she could be completely selfish and devote all her time to developing her career, building strong friendships, and traveling to Europe or Hong Kong several times a year. Hannah had a completely different experience than some of her friends who had married right out of college. She got the chance to discover who she was first, before sharing her life with another person. It seemed a shame that she hadn't appreciated that time then; instead she was hoping for something else.

But if she had been too happy with what she had, would she have been motivated to go after something else? It was much easier to celebrate her single life when she had someone waiting in the wings.

Nate had given her so many hints about his long-term intentions that she didn't mind their gradual pace. During the first year, when they discovered that their mothers both hailed from the town of Snohomish, northeast of Seattle, Nate joked, "This means we should be together. We should just get married."

"Sounds like a plan," she responded playfully. Her heart soared when she heard those words. It was the first time that anyone had even joked about marrying her. During another date, he recounted a recent comment by a friend: "Well, you've finally landed on the one you should marry." Nate didn't share his response, but Hannah didn't think he needed to. The mere mention of it was enough to reassure her that their relationship was headed in that direction. She believed that Nate liked the idea of marriage, even though his own had ended in divorce. He said he had tried hard to be a good husband and father, but he hadn't known his wife was unhappy and was blindsided when she announced she wanted out. He was vigilant to make sure he didn't repeat the same mistakes with Hannah and always asked her, "Is everything going okay? Are we okay? Are you happy?"

And Hannah was.

But she had begun to feel a new biological pressure. She was forty-two and wanted to be married first before trying to become pregnant. Even if she had a baby in the next few years, she would be close to retirement when the child went to college. She also wanted to make sure she and Nate had a chance to enjoy their golden years. His kids would be in college soon, and he had friends who already had finished raising their children and now had time to travel. If she was going to ask him to start again with a new family, she felt she should be sensitive to his timing too. Dr. Winslow had given her the deadline of age thirty-nine to freeze her eggs; she gave herself a second deadline of forty-five to thaw them.

However, her plan had a possible hitch: she still wasn't sure Nate wanted another child.

Since their short discussion on New Year's Eve a few years earlier, Hannah had not brought up the subject again. She relied on hints, except this time they were hers. When friends had asked her in front of Nate if she wanted kids, she responded, "I definitely want to try." She felt as if she had made her desires clear, and she interpreted Nate's silence as a sign that he didn't object to the idea. If he really didn't want more kids, wouldn't he have told her by now?

For someone who had gone to such lengths to safeguard her ability to have a baby, it didn't make sense that she would take such a dangerous gamble regarding who would be the father. But Hannah's approach of "If I don't say anything, he can't say no" had a strange logic. Not only did it buy her more time to deepen their bond, but to actually push Nate for a decision was, in her mind, to invite scrutiny and mess with a good thing. Sometimes she couldn't believe that she had come this far in a relationship. She was *so* close to getting everything she wanted. She preferred to stay in the realm of wishful thinking and simply shut out her doubt. She knew it wasn't the healthiest approach, but given the importance of his answer, it was one she could handle. Nice people got married. Didn't they have children too? Maybe they simply expected to.

But occasionally Hannah felt waves of panic that stopped short of a full attack. "What if he did say no?" she asked herself. She couldn't fathom starting over with someone new, and she definitely didn't want to be a single mom. She saw how much Nate's kids needed their dad.

So one evening Hannah summoned the courage to say the four words hated by men everywhere: "We need to talk." They talked, and she learned why Nate had been ignoring the topic. He *did* have reservations about getting married again: he and his ex-wife had angry arguments, and he hated the way Hannah sometimes exploded at him. His ex had complained when he spent time with his guy friends, and he worried that Hannah would do the same. As for another baby, he confessed he wasn't worried about himself; he was worried about Hannah. "Given what you've told me about your panic attacks, how do you know you won't get postpartum depression?"

Hannah immediately felt defensive. But she quickly rallied with a rebuttal: she promised she would tell him when something was bothering her rather than keep her feelings bottled up. She also promised they would maintain the freedom they had now to pursue their own hobbies or go on separate vacations, including his annual hunting trip. She reassured him that although she still struggled with anxiety, she hadn't had a full-fledged attack for several years. "You don't need to worry about depression," she insisted. "I know all the signs. If I needed to, I would get help." She was surprised this was his concern, considering that terrible period in her life had ended five years earlier.

More difficult conversations followed. Then Hannah asked Nate if he wanted to look at engagement rings. He said, "Okay."

The following Saturday, Nate left early with his kids and told Hannah to drive to an exit on Interstate 90 and call him for further directions to a surprise location. Hannah had guessed he was planning to propose and brought gifts for the kids—a Swiss Army knife for Jason and a pair of earrings for Holly—to make them feel included. This was not how she had pictured the man of her dreams asking her to marry him. She had imagined a fancy restaurant, not a hot and dusty picnic table in the mountains. But she realized this was true Nate. Having a picnic with his kids by the river was his idea of a special occasion, and he was asking her to share it with him.

When she arrived, she welled up as she saw Champagne glasses and a dozen roses on the picnic table. On cue, the kids scampered, and Nate got down on one knee and asked, "Hannah, will you marry me?"

"Yes, Nate, I will," she responded, her face wet with tears.

• • •

In August 2008, a year after getting engaged, Hannah and Nate were married in an old farmhouse in historic Snohomish. The day came together exactly as she had planned and was full of special details, such as her elegant spaghetti-strap gown with lace overlay and the twenty-five fruit pies her mom and mother-in-law had made for dessert. Hannah couldn't stop grinning as she and Nate exchanged vows, in a lush garden before dancing to Etta James's "At Last." Hannah didn't care if it was a cliché; the song said everything.

Nearly two decades after meeting as awkward, insecure college students, she and Nate were finally united as husband and wife. Hannah believed they belonged together. They just had to walk down different paths to know that.

Despite her jubilation, the first months of marriage weren't easy. Hannah had a hard time leaving her beloved house and moving into Nate's place. After years of following a carefully constructed schedule of date nights and alternate weekends off, she had to adjust to living with him full time and his kids part time. The transition became harder when Nate decided to completely remodel his 1980s split-level ranch house a month before their wedding. He tore up the floors, ripped out the popcorn ceiling, and put in new windows. Four months after their wedding, he finally put down carpeting, and Hannah moved in.

Another challenge was adjusting to the role of stepmom. Hannah had gone out of her way not to intrude on the kids' time with their dad, so much so that she wasn't sure how to blend in. Nate and his kids had been used to being a threesome for so many years that Hannah sensed they sometimes missed the days when it was just them. Often, when his kids were over, they would make plans for the weekend and then invite Hannah along as a courtesy. She wanted to feel like a real member of the family and asked Nate to consult with her rather than simply inform her of the schedule.

She also couldn't stand the way Nate and his kids let the house degenerate into an awful mess before performing a marathon cleanup. At first, she wanted to ensure a smooth transition and took it upon herself to keep the house tidy, but she quickly grew tired of having to nag them to put their dishes in the dishwasher or take their belongings to their rooms. The kids obeyed, but Hannah wished she could return to her role of weekend fun person. Nate, protective of his kids, developed the habit of running interference for Hannah whenever she had an issue with one of them. She appreciated his support, but she felt his meddling prevented her from establishing her own way of dealing with conflict with them. Slowly they started to form their own relationships. She drove Holly to school in the mornings, and gradually the girl opened up about her classes, friends, and love life. She and Jason chatted about their favorite bands, U2, Guns N' Roses, and ACDC. When Hannah went grocery shopping, she made sure to buy chocolate syrup for Jason's milk or the makings for meatloaf, a family favorite. One evening when Nate had to work late, she took the kids to TGI Fridays for dinner. When they returned home, they put on an impromptu show for her of camp songs and dance routines. For Hannah, it meant they had accepted her as a part of the family.

A few months after their wedding, Hannah was ready to try to get pregnant. She wished she could spend a few more years enjoying being married, but at nearly forty-four, she couldn't wait any longer. So she started monitoring her ovulation schedule and made sure they had sex when she was most likely to be fertile. She started reminding him, "If this doesn't work, you know I always have those eggs in Florida." She knew Nate was a deliberate plodder who needed to get used to ideas, so she wanted to give him plenty of advance notice.

After six months of trying naturally with no success, Hannah booked an appointment with a reproductive endocrinologist, who was recommended to her by her ob-gyn. She wanted to inquire about having her frozen eggs shipped to Seattle, but the elderly doctor refused to give her hope that the method would work. "Let's be realistic," he said. "You're forty-four. Your chances are pretty slim. Even going to Florida is pretty slim."

Hannah wasn't surprised when he estimated that her chances of getting pregnant naturally at her age were less than 5 percent, even though her hormone levels were still in the normal range. She wondered, though, how he could comment on her frozen eggs when he seemed to know so little about the procedure. She didn't bother to ask about IVF, since she figured those success rates were abysmal too, and she didn't want to waste the money when she had her eggs waiting for her across the country. Although she had little confidence in his opinion on frozen eggs, she left the appointment in tears. She had been counting on those eggs for six years.

When Nate saw Hannah crying, he consoled her, saying, "Where there's a will, there's a way. Let's give your frozen eggs a shot." Hannah wiped away her smeared mascara and beamed.

She had known all along that her frozen eggs might not work and had tried to prepare herself over the years to accept the worst-case scenario: that she wouldn't have a baby. She knew she shouldn't get her hopes up, even though she figured she had at least a couple of shots with twenty-one eggs.

To cope, she compiled a list of the advantages of not having kids. For years, she had watched her married friends who didn't have kids enjoy their free time traveling, camping, or playing cards, and she wanted time to do the same with Nate. She even considered abandoning her career in fashion (she was sick of the industry's egos and personalities) and going back to school to become a pharmacist. She reassured herself that not having biological kids didn't mean she wouldn't have a family. She had a devoted husband, and she could work on building closer relationships with Holly and Jason. "You're still going to be happy," she told herself.

Sometimes the list felt like a lousy consolation prize. At other times she had to admit to herself that she had mixed feelings about having a baby. She feared that a child would change her relationship with Nate. She worried about losing her old friends, who were now planning fun trips she would miss. She didn't think she could relate to younger mothers, who would make up her new social circle. Then there was her free time. For nearly three decades Hannah had been used to going to the gym or seeing a movie whenever she wanted. How would she handle being tethered to an infant during most of her waking hours? She even nurtured a small hope that her frozen eggs wouldn't work. That way, she could say, "Well, I tried. I did everything I could to have a baby. I've done my duty as a woman." Then she'd be off the hook.

In November 2008, nearly a year and a half after getting married, Hannah and Nate flew to Florida to try to have a baby. They had completed all their preliminary tests in Seattle, including screens for sexually transmitted diseases and an evaluation of Nate's sperm count. In the meantime, Hannah wore hormone patches for several weeks to thicken her uterine lining to prepare it to receive an embryo.

During those few months, Hannah fell in love with her husband all over again. She knew this trip wasn't easy for him. The conception of his first two kids had been so uncomplicated; he had probably never anticipated giving a sperm sample in a tiny closet in a nondescript hospital office suite clear across the country for his third.

Dr. Yang thawed eight eggs, and five turned into thriving embryos. Hannah was relieved she still had thirteen eggs left for a couple more tries. Dr. Winslow didn't want to transfer all five for fear of creating multiples, so he chose the three best-shaped embryos and deposited them into Hannah's womb. "Don't get too excited," she told herself as she curled up on the transfer bed afterward, employing her old defense mechanism of avoiding disappointment. "It's not going to work."

Sarah

When the Bologna conference was over, I invited Dr. Debra Gook, who in the early 1990s was the first scientist to prove that freezing did not damage human eggs, and Dr. Thomas Toth, who had started an IVF program at Massachusetts General Hospital in Boston, to lunch. After a lazy meal of wild boar pasta and red wine in Bologna's old city, we strolled under the medieval porticoes admiring the Christmas decorations. The conversation turned to why I had frozen my eggs. Dr. Toth, a father of three, urged me not to rely on them. "Don't wait too much longer to start your family," he said.

"Well, what if I freeze a lot of eggs to improve my odds?" I asked them, feeling like a teenager trying to negotiate with her parents.

"With the technology improving and lots of eggs, I guess that's possible," conceded Dr. Gook, who had two children. But she pointed out that the argument could be misleading. "You have to thaw the same way you froze," she said, meaning I was locked into an operating system; if my eggs had been frozen with a slow-freezing method, they had to be thawed using that same method. I could still benefit from improvements in how the eggs were fertilized and how embryos were grown in the lab, but the two most important parts that determined the success of egg freezing—the methods of freezing and thawing—were done deals.

Toth jumped in. "You're putting your hands in a lab," he said. "Putting all that trust in technology is a lose-lose, whether you're freezing five or forty eggs. You may be one of those people who don't get pregnant. People always think they'll be in the majority. They always think they'll be in the 98 percent. But we see the 2 percent." In other words, they saw disappointment I couldn't even conceive of.

Then Toth came to the heart of the matter. "I don't understand why you would freeze your eggs if you have a boyfriend," he said. "I tell my patients, 'Figure out the relationship first. Then have a baby.' If I was the guy, I'd feel uncomfortable if my girlfriend got her eggs frozen. It would mean she wasn't serious about me." Gook nodded in agreement.

"Not serious?" I protested, taken off-guard. "What if you did it to take the pressure off a relationship because it was too early? To give the relationship a chance?"

"Waiting for men to make up their mind will never happen," Toth said definitively. "Women drive the decision."

I suddenly realized that apart from Paul and my doctor, this was the first male perspective I had heard on the topic. I had thought that my decision to freeze my eggs sent a message that I was serious about having children, but I could see Toth's point that it also communicated I wasn't serious about having them *now* and had given myself a backup to have babies with someone else. I wasn't driving *anything*. Egg freezing allowed ambivalent men—and women—to put off the decision indefinitely. There was no biological imperative to force *anyone's* hand.

I suddenly felt nauseous, a state not helped by my red wine buzz and lingering jet lag. I had effectively removed any leverage I had to resolve the baby issue and move the relationship forward. Toth made it sound so easy: the woman put her foot down, and the man gave in—if they didn't have an accident in the meantime. That was how it had happened for centuries.

Paul said he was grateful I had frozen my eggs, but I did wonder if the act bothered him in some way. I couldn't think of a modern equivalent—perhaps agreeing to an exclusive relationship while keeping up a Match.com profile? Whatever the case, egg freezing was not a gesture of confidence in our future.

That evening, I returned to my tiny Italian hotel, climbed into the tiny Italian shower, and let the water run over me. The pasta felt heavy in my stomach, and my head started to spin. I poured myself some of the local Sangiovese wine that I had bought at the grocery store and drank a few sips, but it made me feel worse.

Paul had sent me a sweet email wishing me well on my work and telling me how much he missed me, but I didn't feel like answering it. I was in bed by 7 p.m.

Paul and I still persevered through our summits, but we were no closer to an agreement on whether to have children. One week, we read a chapter together on the importance of posterity, but that didn't hit home for either of us. The conversations returned to how we would manage a baby. What would our Sundays look like? I suggested that we could jog with her in the park, take her to brunch, make love during her naptime, and put her to bed by 7 p.m. "Our lives wouldn't stop," I insisted. We could still do the things we loved. We just had to plan ahead more.

But the conversations always came down to this: children change your life, and there's no guarantee it will be for the better. In fact it could be for a lot worse. What if our son was a drug addict? What if our daughter had expensive special needs? Why would we give up what we have now, what we know is working, for something so unpredictable?

"Why don't your fantasies ever include the hard stuff?" he asked, pointing out that while I had been admiring the family biking together in Bar Harbor, he saw their kids screaming and running around the hotel lobby during checkout. "Relationships require work. Kids add another level of complexity. By definition, we would have less energy to focus on each other. I want to spend my time making you happy." I wanted to tell Paul he was naïve to think he could be my everything; children provided a different kind of love that was also important. I think he believed that if he just tried hard enough to be a good boyfriend, he could override the ache in my gut. I was touched and baffled at the same time.

I insisted that we'd protect our time together by hiring help or trading babysitting with other families. We would plan for the hardships. We would educate ourselves about the challenges. We had to take risks to enjoy the richness of a full life. I told him of an often-quoted Gallup poll in which two-thirds of childless people over forty said they wished they had kids, and the majority said they would have had at least one child if they could rewind their lives. Only 4 percent said they didn't plan on ever having kids or were glad they never had any. He replied that he'd take a look at that study.

Real life didn't help my case. If Paul and I walked through Central Park, we inevitably came across a child throwing a temper tantrum while the parents bickered. If we went out to brunch, we sat next to parents shoveling their food while their squirrely toddler tried to run around the restaurant. If we traveled, we picked the airport security checkpoint in which the baby was having a meltdown while the poor mother had to deal with folding a stroller, fishing out her laptop, and taking off her shoes. Paul wasn't the only one cringing; I was too. Perhaps I was more ambivalent than I realized.

In the meantime, I thought up new supporting arguments. We could have just one child. I had come across another study concluding that mothers experienced a happiness gain of 20 percent after the birth of their first child, though that gain dropped slightly with additional kids. Men experienced a smaller gain, but it was still a win-win. "So what about having just one little baby?" I inquired. "Maybe that would be the perfect compromise."

He replied that the big decision was whether to *have* kids or not. It didn't seem as big of a leap to have a second.

I said we could postpone having children. We could make a list of all the places we wanted to go and things we wanted to accomplish before having children and start working our way down. If we needed more time, maybe we could look into an additional insurance policy by freezing some embryos made from his sperm and eggs from my next harvest. The practice had a slightly higher success rate than egg freezing. But I knew I was reaching. Stashing away another form of reproductive matter was no substitute for a decision.

Once, I curled up on his couch in the fetal position and cried, "But I just really, really want one." I had been reduced to begging, and I feared I had failed the most important lesson of every relationship advice book: "Ladies, if he does not want what you want, get out and find someone who does." I was starting to humiliate myself. Worse, I was wasting time.

• • •

I focused on putting away more eggs. At this point, all I had were six eggs, one 30 percent shot, which had cost me $16,000. I needed a frequent-freezer plan. I had read about a Canadian fertility expert named Dr. Seang-Lin Tan of McGill University in Montreal who already had nearly thirty frozen-egg babies to his credit. My heart skipped a beat as I researched his clinic to see if egg freezing would be cheaper in Canada. It was! The national health care system offered foreigners a package deal of three rounds of egg freezing for $12,000, less than it cost for one round in New York, and was a clear reflection on the value of banking in bulk. The hitch: I would have to pay for early monitoring appointments at home before flying to Montreal for the final few days before each retrieval. Still, the total cost of each round was less than half the price of my first, and I was thrilled when my parents offered to cover it. Best of all, Dr. Tan said he would try a new drug regimen that might significantly boost my egg count. I hoped to produce enough eggs to have a real safety net, which Tan defined as approximately thirty eggs.

His logic: If thirty eggs resulted in ten good embryos, and those were transferred two at a time—three at a time for women who were over thirty-five when they froze—I'd have several attempts, with a 40 percent chance during each attempt of at least one embryo implanting and resulting in a live birth. If one batch of thawed eggs gave me more embryos than expected, the extra embryos could be frozen again: the "double freeze" I had heard about at the conference in Bologna. All that math meant that thirty eggs would be good for one baby and maybe two.

I also liked the added precaution of storing my eggs in two different countries. During my road trip with Paul, I had made a visit to the New England Cryogenic Center to see where my eggs were kept. Extend Fertility reps liked to point out that the facility is located in the Boston suburb of Newton, which, in 2005, was deemed the safest city in America, according to FBI crime statistics. It was also far away from earthquake fault lines, terrorist targets, and tornado alleys. The director showed off the extensive security system, the backup generator, and the machine that automatically replenished the liquid nitrogen daily in forty or so tanks that held umbilical cord blood, stem cells, bone marrow, embryos, sperm, and eggs. In case of Armageddon, I was told, the tanks could last a month before needing to be refilled. Finally, he showed me where my eggs were kept, and I was amazed to see them stored with eggs from Extend's two hundred or so other clients in a stainless steel canister that was no bigger than a helium tank at a party goods store. The director opened the top and reached his hand into the vapors to show me how all of the egg-containing straws had been loaded into holders called canes and were sitting in metal goblets at minus 196 degrees Celsius.

I couldn't get over the fact that so many women's hopes and dreams were suspended in such a small tank in a nondescript brick building near a dry cleaner and a Jiffy Lube. Part of me liked that we were all in there together, supporting each other in some sort of subzero sisterhood. But it was surreal to know that our treasured genetic material was guarded by strangers who were entrusted to keep them perfectly intact for us until we were ready to use them. The multiple security systems were reassuring, but our little eggs still seemed so vulnerable—and so far away.

Now I was planning to deposit another batch in a foreign country, which gave me even more separation anxiety. I was proud of myself for sourcing such a deal, but I quickly learned there was a reason Americans weren't eager to take advantage of the Canadian socialized health care system. The medical expertise was excellent, and the staff was warm and responsive, but the task of navigating two unwieldy medical systems turned out to be a huge pain. My local lab wouldn't accept my Canadian order form for blood tests, so I called my ob-gyn and begged her to fax one on the spot. My pharmacy wouldn't take Canadian prescriptions, so I also asked her to write me a new set. One drug had to be custom-mixed, since the American version came in premeasured packs, an adjustment requiring several more phone calls.

I paid hundreds more for three visits to a fertility doctor who agreed to do monitoring for Dr. Tan. His office was located in Brooklyn, a forty-five-minute subway ride away. For my first appointment, I walked slowly from the station to the office. I had been taking a longer drug protocol that gives the smaller eggs a chance to catch up to the faster-growing ones and was supposed to result in a better harvest. As I hoisted myself onto the examining table, I dreaded hearing the news that I had produced another paltry batch.

But as I watched the ultrasound root around my ovaries, I listened to her counting. "Nine, ten, eleven, twelve on this side."

"How many total?" I asked.

"Oh, around twenty or so, depending on how they all grow," she answered. "Don't worry! You've got lots."

Lots! I had lots! I loved hearing that word linked to my egg production. Even if only half resulted in mature eggs, I would be one-third of the way to Tan's target of thirty, with two more cycles left.

I was so happy with my follicle count that I didn't even mind spending extra nights in Montreal so Tan could wait for the right date to harvest them, despite the fact that airlines do not accept egg freezing as a valid reason to drop penalties for last-minute ticket changes.

When I arrived at the castle-like Royal Victoria Hospital for an ultrasound, I was thrilled when the examining doctor looked at the monitor and announced that fourteen were still growing strong. However, I was shocked to learn that "for liability reasons," Americans weren't placed under general anesthesia for the retrieval surgery. Instead I would be given a local anesthetic and some sedatives. "You'll be fine," the nurse reassured me.

The drug combo was supposed to make the retrieval bearable, but because I have low blood pressure, the nurse held off on the sedatives for fear I'd stop breathing. I felt the needle each time it pierced an ovary to suck an egg from a follicle. "Give me a little cough," the retrieval doctor instructed, hoping to distract me so she could punch the probe in. But I kept jerking, and she quickly grew frustrated. "You've got to hold still," she scolded.

Tears began streaming down my face, and by the time she had finished one side, I was sobbing. The heaving made her job even harder. "Do you want me to stop?" she asked, her nervous dark eyes peering over her blue mask. I don't usually cry when I'm in physical pain, but the agony had unleashed a torrent of intense sadness coupled with gratitude.

I felt a primal urge to fight for these eggs I had worked so hard to grow. I had spent years indulging my ambivalence, engaging in long theoretical discussions, and breezily chatting about endless options. Now I was lying naked on a cold operating table in a foreign country while a stranger put a needle up my hoo-ha to try to save—egg by precious egg—the last of my fertility. During moments like these, you own your desire to be a mother.

"No," I replied. "Get them all!"

By that time, my blood pressure had gone up, so the nurse added the sedatives to the IV. I could still feel the pinching, but the sensation was tolerable now. My body relaxed, and I wiped away the tears and sank into the buzz. On the monitor I watched an embryologist fish out the pearly eggs from the ovarian flotsam of blood and tissue.

"Fourteen!" the nurse announced when it was all over. "We got all fourteen." I beamed through my drug haze as I calculated that, combined with my New York stash, my count was now up to twenty.

The fee for three cycles of egg freezing at McGill included a phone consultation with the clinic psychologist, Dr. Janet Takefman, but the first available appointment wasn't until several months *after* the first retrieval. I had welcomed the chance to talk to another therapist about egg freezing, but by the time the appointment arrived, the last thing I wanted to do was to explain to another stranger how I had ended up in this pickle of falling in love with someone who didn't know if he wanted kids.

I was still smarting from the most recent clash with my mother. "What's going on with Paul? What is he waiting for?" she had asked. I hadn't wanted to explain the summits to her and just told her we were dealing with the issue. "Dealing with what?" she had shouted, tears welling in her eyes. "My God! You've had to freeze your eggs!"

I didn't want to add Dr. Takefman's voice to the rising chorus from my friends and family of "You *have* to listen to what he is telling you and make some hard choices" or "What makes you think you can change his mind?" But Takefman praised me for freezing. She also said that while I shouldn't think of the procedure as an insurance policy, it was valuable for giving me peace of mind and allowing me to move forward with my life goals while working on my relationship.

When I told her about the frustrating summits and how we had resorted to pushing our positions on each other, she responded, "You can stop the marathon talks, which in my opinion go nowhere. As long as you know he's thinking about the subject, that's enough."

I was relieved when she said I owed it to myself to see if we could resolve our differences. However, she warned me against giving up children. "Don't negotiate away a need," she said. "No relationship could survive that kind of regret." If Paul was worried about losing certain freedoms, she suggested making agreements and even drawing up contracts to protect what was important to him, such as going to dinner by ourselves twice a week. I loved her advice but wished our negotiations could be so simple. Paul still had to agree to have a kid in the first place.

"How long do I give it?" I asked.

"I'd wait a year because you'd have to start all over or do it alone," she said. "But don't tell him that." I was surprised she actually advocated taking *more* time, since I assumed she would urge me to break up with him and find someone who shared my goals. I was encouraged that even a mental health professional recognized my relationship with Paul was worth fighting for.

Monica

Five years after clutching her chest in a Minneapolis conference room, Monica still couldn't escape the pains that gripped her chest every few months around the time of her period. She had undergone nearly every major medical test doctors could think of to determine the cause of her pain, but no one knew what was wrong with her.

A routine blood test during her annual checkup finally revealed what her body had been trying to tell her for all those years. Her results showed a high level of creatinine, a waste product that can indicate kidney problems. Her family doctor referred her to a urologist, who found a growth blocking the ureter, the tube that connects the kidney and bladder. He removed the tissue and implanted a stent to open up the tube.

At age forty-one, Monica was diagnosed with endometriosis, a condition affecting 5 to 10 percent of women in which tissue from the lining of the uterus escapes and grows on the ovaries, fallopian tubes, uterine surface, vagina, and other parts of the body. Her other doctors had missed the diagnosis because, like many women, Monica didn't have any of the common symptoms, such as chronic pelvic pain, intense cramps, painful sex, or extensive bleeding.

The tissue was everywhere. The radiologist said there were so many dark spots indicating endometrial growths on her x-ray it looked as if someone had taken a rifle and shot up her pelvic area.

"You have a severe case," announced her ob-gyn.

"Is this the reason I wasn't getting pregnant?" Monica asked.

"Most likely," she replied, explaining that endometriosis is a common cause of infertility.

As for Monica's chest pains, the doctor suspected that the tissue had settled into her chest cavity and irritated the lining. Unlike endometrial tissue in the uterus that normally builds up, breaks down, and is eliminated during menstruation, the tissue in her chest went through a similar monthly shedding process but couldn't leave the body. That's why Monica experienced the pain during her periods.

Her ob-gyn advised she start a six-month course of Lupron, an injected hormone that would reduce her ability to produce estrogen and would starve the endometrial growths. "It will put you into a temporary state of menopause," she said. "If that's not successful, we'll have to treat it surgically." Monica feared it was impossible to individually remove so many growths and that "treat it surgically" meant a hysterectomy to remove her uterus, which is often a last resort in severe cases. If left untreated, endometriosis could spread to her bowels, lungs, and other organs.

She felt nauseous as she trudged to the parking lot. She tried to put on a brave face for her parents, who had accompanied her to her appointment, but as soon as they drove away, she dissolved into tears. Although she was relieved to finally learn what was wrong with her, and why she had been unable to get pregnant with Adam, she was devastated that her condition was so serious and that the treatments were so extreme. At best, she would experience menopausal side effects: hot flashes, weight gain, mood swings, and loss of libido. At worst, her womb would be removed, and she would never be able to carry a child.

Later than night, as she tried to digest the news, she sat at her laptop and looked up stories about women with endometriosis who had become pregnant. Her doctor had explained that her fertility likely would return to normal once she stopped taking Lupron, and Monica was heartened when she read that *Top Chef* host Padma Lakshmi had a baby after undergoing numerous laparoscopic surgeries to remove her growths. Monica also found accounts of women with the condition who had conceived on their own or undergone successful IVF with their own or donor eggs. But there was a catch: even if the Lupron worked, there was always a chance the growths could reappear years later, especially in women with severe cases.

Although endometriosis is estimated to account for some 15 percent of women's infertility cases, it's not always clear how the condition hurts reproductive functioning. There are several theories: If a woman has a thick layer of growths on her ovaries, an egg could have a hard time making its way out of its follicle after being released during ovulation. If a growth makes her ovaries or fallopian tubes adhere to the lining of her pelvis, an egg might not be in the right position to drop into her fallopian tubes. The endometrial tissue could also scar the pelvis, block the fallopian tubes, or stimulate the production of cells that attack sperm.

Even if Monica bypassed these problems by having embryos made from her frozen eggs transferred directly to her uterus, she faced another challenge. Some doctors speculated that endometrial tissue might lead to the overproduction of hormones that could interfere with embryo implantation. Monica was suddenly grateful she hadn't wasted her frozen eggs by unknowingly using them in a sick body.

All these years, she had been counting on her frozen eggs as a backup for old eggs. She had never imagined they would be a backup for illness too.

As Monica prepared to start her course of Lupron, she was terrified of becoming one of the "crazy menopausal women" she had read about. *Menopause* was almost as unseemly a word as *infertile*, and soon she would be both. Even though both states were likely temporary, she was disturbed that these terms could so thoroughly rattle her sense of female identity.

The side effects came on slowly. She attended weekly acupuncture sessions and swallowed black cohosh pills, an herb that menopausal women use to ward off hot flashes. At first the supplements appeared to work. During the first month, Monica only occasionally suffered hot flashes. But by the second month, she routinely woke up three or four times in the middle of the night drenched in sweat. During the day, the hot flashes came on quickly; within ten seconds of feeling a tingle, she became so hot she sought immediate relief by blasting the air conditioner or splashing water on her face. She was either unable to stay up past 9:30 p.m. or couldn't get to sleep at all. She spent her forty-second birthday that summer feeling bloated and sluggish. She stopped having a period.

Even though she didn't feel like going out much, she still tried to date. But she didn't feel the usual pressure to screen men for their willingness to have children. There was the obvious issue of not knowing whether she would be able to have a baby, even with her frozen eggs. She knew she could always have them fertilized with her partner's sperm and carried by a surrogate—her sister-in-law had even volunteered for the job—but that option seemed complicated and expensive.

The other issue: she had started to doubt her own desires. Monica loved being an aunt but was exhausted after spending time with her four-year-old nephew and two-year-old niece. Sometimes she wondered if stepchildren could be enough. She had been surprised by how much she enjoyed getting to know the three teenage boys of a forty-four-year-old man she had briefly dated, but she was relieved when they returned to their mother and the house became quiet again.

Monica revised her Match profile. In response to the question whether she wanted children, she changed her response from "Someday" to "Not sure" to deter men who were eager to have children. She wanted to leave open the option, since she wasn't sure if she'd really had a change of heart or if her illness had put her in a different headspace. Whatever the case, the notion of having to consider someone else's baby desires suddenly felt overwhelming. So she made a point of asking her dates up front, "How serious are you about having kids?" Then she delivered her prepared line: "I'm open to having them, but I need to let you know it might not be possible to have my own." It was a far cry from her thirty-seven-year-old self with her spreadsheet. Her forty-two-year-old self had to focus on her health.

In addition to the Lupron treatment, Monica endured four surgeries to fix her kidney: one to remove the endometrial growth and insert a stent into her ureter; one to replace the stent with a smaller model after her urologist suspected it wasn't working; another to replace it again after she showed no improvement; and another to remove the stent and resect her damaged ureter. Her ob-gyn attended the final surgery and took the opportunity to check Monica's pelvic area for endometrial growths with a scope through the surgeon's laparoscopic incisions. She couldn't find any.

"The Lupron worked," her ob-gyn triumphantly announced. "I'll want to recheck it in six months, but it appears you have a clean bill of health."

Monica felt a deep peace settle in her body. As soon as she stopped taking the Lupron, her hot flashes disappeared and her bloated stomach flattened. She felt her energy return, as if she had emerged from a long flu.

Kelly

Kelly was waiting in a Raleigh hotel conference room to give a slide show to her volunteer group when she heard her phone vibrate. Her stomach lurched when she saw the area code for Jacksonville: the clinic was calling with the results of her pregnancy test. She was up next, but she quietly slipped out into the hallway to answer her phone. The nurse's solemn voice told her everything she needed to know. "Kelly, we got the results back, and they're negative," she said. "I'm so sorry. You can stop your medication." Kelly thanked her and immediately called Dan.

"Are you okay?" he asked carefully. She and Dan had been married for only four months and had never gone through something like this together. "Honey, I'm fine," she insisted, trying to sound upbeat. She had no time to dwell on the news; she had to start her presentation. As she made her speech, she thought about how strange it was to be showing photos of smiling volunteers during a recent retreat when she felt so numb.

She tried to come to terms with her disappointment by reminding herself that it was only her first try. She knew that IVF patients routinely went through several treatments before getting pregnant. But unlike IVF patients, who often have to wait a couple months between stimulation cycles to give their ovaries a rest, Kelly had already done the hard work of producing eggs and could undergo another transfer in two weeks, during her body's normal ovulation cycle. Dr. Yang would thaw the remaining ten eggs and inseminate them with the rest of Dan's frozen sperm. She and Dan would then travel to Florida when the embryos were ready to be placed into her womb.

Kelly had earmarked those eggs for a second child, but she tried to remain grateful that she still had one chance left.

When they arrived in Jacksonville, she was ecstatic to learn that Yang had thawed out only five eggs, meaning she still had five left for another try. Plus, they hadn't used all of Dan's frozen sperm. But the best news of all was that four had successfully fertilized and developed into thriving embryos. Kelly signed a new consent form, and Dr. Winslow explained there was less than a 1 percent chance of all four implanting.

Later that evening, she and Dan speculated on what it would be like to have multiples; with four embryos, they figured it was reasonable to expect her to become pregnant with at least twins or triplets (a 20 percent and 7 percent chance, respectively). They let their imaginations go wild with all of the possible combinations—even quads. What would they do with all those babies? Would they need a bigger SUV? Would they ask family to help, or hire the twin's former nanny? They'd have to knock out a wall above the living room and add a couple of bedrooms in the attic space. Dan's mind raced with construction timelines. They'd have less than nine months to get everything done.

In the weeks that followed, the pinching sensations and PMS symptoms felt the same as they had after her first transfer. But this time Kelly believed the nurse's dismissal that they didn't mean anything; both she and Dan were too excited to doubt her. On Thanksgiving Day she sent out an email to close friends and family members to announce the news. "Your prayers have worked . . . at least in the world of quantity!" she wrote. "Yes, all four were transferred and as overwhelming as it may be, Dan and I welcome all four into our lives!"

She was flooded with voice mails and emails of congratulations. Her friends and family were just as confident.

A couple of weeks before Christmas, on the day Dr. Winslow's office would call with the results of her pregnancy test, Kelly checked her cell phone several times to make sure it was charged. She carried it with her when she walked upstairs or went out into the yard. She did not want to miss this call.

She didn't want to jinx her chances by outright expecting good news, but she didn't want to give into doubt either. Still, she hadn't imagined how she and Dan would celebrate that evening.

She answered the phone by the second ring. As soon as the nurse said, "Hi Kelly," in a business-like tone, Kelly's body went rigid. She had been sure at least one embryo would have made it.

She spent the next couple of days walking around the house in a daze, looking for any distraction from her anguish. She had lots of sex. She binged on greasy meals of burgers and fries. She drank red wine, which had been forbidden before and after the transfer.

She dreaded writing the email update. What was she supposed to say? Should she be straightforward? Or tell them the truth: that her heart was breaking? The last thing she wanted to do was prompt a bunch of phone calls from friends and family. She couldn't make sense of her own disappointment, let alone provide perspective for others. She decided it was best to explain that she had one last shot. "Hello to all and thank you for your continued prayers," she wrote. "Unfortunately, the pregnancy test was negative again. We were certainly disappointed and even a little angry; however, we continue to remain hopeful that we will indeed be blessed with children . . . one way or another! We look forward to starting the process again in January."

Kelly hadn't wanted to wait until January, but the clinic would be closed during the holidays. She usually liked to keep bad news at bay for as long as possible, but she wanted to know the results now. Despite what she had written in her email, she wasn't hopeful about her remaining five eggs. Of the fourteen eggs she had thawed so far, seven embryos had developed in the lab. But all of them had stopped growing inside her body. Why would the results be any different for these last five eggs? Clearly, something wasn't working.

In the meantime, she tried to get excited about her first Christmas with Dan as newlyweds. She kept busy with parties, decorating, and volunteering. But she couldn't stop thinking about what could have gone wrong with the past two transfers. She wondered why Yang had successfully created four embryos out of five eggs during the second round compared to just three from nine eggs the first time. Did other staff actually handle the eggs? If so, were different people working on those days? Did they use different methods? How did they choose which eggs to thaw?

Even if the success of her embryos depended on the talent of her embryologist, that wouldn't explain why her body had somehow failed to nurture them. Winslow had conducted tests to make sure her uterus was in good shape to receive them. But Kelly wondered if there was something else wrong with her that couldn't be diagnosed. Did she have an incompatible uterus or hormonal imbalance? Had she wasted her eggs?

But Kelly didn't ask Winslow. She assumed he would reassure her there was nothing more that could be done to improve her chances. Her questions wouldn't change the outcome anyway. It was in God's hands now.

The result from the final batch was devastating; out of five eggs, only one embryo developed. Kelly tried to be optimistic in front of Dan, saying, "This has to be the baby. There's no way we can go through all nineteen and not have a pregnancy." She understood intellectually that she had other options to have a child, but she had grown attached to those eggs that had been waiting for her for so many years. Now she was down to one.

Kelly coached herself to prepare for a negative result: "You are so happy right now, even without kids. If this doesn't work, you'll be heartbroken. But you'll be okay. God has a plan for you." But she repeated her pep talk so often that the words lost their power.

Kelly was driving back from returning Christmas presents when Dr. Winslow's nurse called for the last time. The final pregnancy test was negative. After she delivered the news to Dan, her lips quivered, and big tears rolled down her cheeks. Kelly didn't cry very often, and out of habit she tried to stop. But she heard a more compassionate voice inside her that said, "Kelly, you just had your third negative result. Let yourself cry for a minute." Pulling over on an old country highway in the pale winter dusk, her tears gave way to sobs that made her gasp and shook her body. She had never known such intense sorrow.

When she arrived home, Dan ran outside and took her into his arms. She glumly joked that she could have a whole bottle of wine with dinner that night, if she wanted. He responded with a helpless look. His tendency was to deal with grief by fixing things, but there was nothing anyone could do to make her feel better. She couldn't even bring herself to think of all the lost time and money: nearly $30,000 for everything, including Dan's sperm aspirations. She just let herself wrestle with the awful truth. The whole thing was a bust.

Hannah

Hannah had never thought there was such a thing as being a little bit pregnant, but two weeks after having three embryos transferred to her uterus in Florida, that's what she was. "Well, your levels are not what we would hope," the nurse from Dr. Winslow's office told her of the pregnancy hormone, human chorionic gonadotropin (HCG), measured in her blood the day before Thanksgiving. "However, you still could be pregnant." The nurse explained that a slow hormone response was one of the effects of using frozen eggs. Since her body had been artificially prepared with estrogen and progesterone patches, it took longer for her natural hormones to kick in. In pregnant women, HCG doubles every two to three days. "We just need to give it more time," the nurse added, explaining that Hannah would need to be retested the following week.

Hannah assumed the worst. "I knew it wouldn't work," she told herself, searching for explanations: her body was too old to support a pregnancy; her eggs had been frozen too long; it just wasn't meant to be.

She had tried to anticipate that the transfer wouldn't work, but she was surprised at how disappointed she felt. She was heartened that neither her mom nor Nate seemed discouraged. "I guess we'll just wait and see," they both said.

So Hannah settled into a state of suspended animation and made it through multiple Thanksgiving dinners. Several family members knew she and Nate were trying to have a baby, and she was grateful no one asked for a progress report.

After an agonizing four days (the clinic was closed over Thanksgiving weekend), Hannah gave another blood sample to her local clinic, which sent the results to Winslow's office. While she waited for news from Florida, she called the clinic to ask if someone could tell her the results in the meantime. "I know I'm not supposed to do this," the nurse said. "But it's looking great. Your levels have shot up. They're exactly what we want." A nurse from Winslow's office interpreted the numbers: not high enough to declare a pregnancy, but definitely moving in the right direction. Hannah was so thrilled she briefly let a little excitement override her anxiety.

"It's looking better," she excitedly told her husband.

"Sounds good," Nate replied. Hannah cringed when she heard his flat tone. She hoped he was reacting to the uncertainty of the news. But when he said he couldn't take time out of his work schedule to accompany her to the ultrasound a few days later, she wondered if she had made a huge mistake pushing him to have another child.

She told her mom that Nate didn't seem enthusiastic about the possibility that she was pregnant. "Don't worry," her mom said. "It's not real to a man in the beginning. He'll come around." Hannah felt vaguely comforted.

Nearly a week after her first blood test, an ultrasound technician said the words Hannah never thought she would hear: "Congratulations. You're pregnant!"

"Are you sure?" Hannah asked, not knowing what else to say. "And it's just one?"

"Yes, it's just one," the technician said.

Hannah left the office feeling as if she were floating. She fished her cell phone out of her bag and called Nate. "Guess what? I'm pregnant," she announced cautiously.

"Okay," Nate replied. Once again she felt deflated. In the movies, men cry tears of joy when they hear the news. But he simply said, "Are you happy? I know you wanted this."

"Yes, I am," she reassured him. But as the news sank in, her delight turned into trepidation. "Holy shit!" she thought. "The frozen eggs actually worked. Now what do I do?" She'd had such low expectations that it took her a while to comprehend that one had turned into an actual baby.

An actual baby who would be born to a forty-five-year-old who came home from work exhausted and was in a high-risk category for high blood pressure and gestational diabetes. Hannah was careful not to let herself feel overwhelmed and savored the sweetest part of her miracle: regardless of her age, her baby was made from thirty-eight-year-old eggs. That meant the child had a 1 in 175 risk of being born with Down syndrome, compared to 1 in 38 if conceived from Hannah's forty-four-year-old eggs.

Hannah constantly did age calculations: she would be fifty-five when her child was ten; she would be ninety when her daughter was her age. She consoled herself that whatever happened to her or Nate, their child would never be left alone. Her brother had seven kids, and of course her baby would always have Holly and Jason.

Her seventy-year-old mother, who had given birth to her at twenty-five, once again told her not to worry. "It's a blessing to have a child at any age," she said. "You could die in childbirth or be killed tomorrow. At least you've brought a baby into this world and created life."

Although Hannah was part of a growing trend of women having babies at her age, she still had wandered into a lonely demographic, at least in suburban Seattle. She was out of sync with her own peer group, since most of her mom friends already had kids in high school. Even those who started late were nearly a decade ahead of her.

Hannah suddenly felt out of sync with her older single friends, like Lori. The two used to be in the same boat, but now their lives seemed vastly different. Lori had a deep desire to marry and have a family, but she hadn't been in a major relationship for several years. She talked half-heartedly about adopting a baby someday on her own but feared she was getting too old to meet many agencies' age requirements. Hannah could tell she was losing hope and ached for her. During their get-togethers, she focused the conversation on Lori's job as a marketing executive.

Hannah was grateful she could enjoy the camaraderie of other new moms at her work. She managed a team of sixteen, mostly women, and five were expecting at the time she learned she was pregnant. Several more on her floor were pregnant, prompting her male boss to joke,

"Don't drink the water!" The women were mostly in their mid-twenties to mid-thirties. For years Hannah had felt left out as she watched her coworkers get married and pregnant and often wondered what her staff thought of her. Did they think she didn't want kids? Or did they assume she was too old to have them?

She never imagined how much fun she would have revealing to her coworkers how she conceived her baby. "Remember when I went to Florida?" she asked, thoroughly relieved to share her secret. She guessed some people had been puzzled about why she suddenly left for a sup- posed family reunion right before a big sales meeting in early November. "Well, a long time ago, I froze my eggs," she explained. She didn't get the sense many people understood the procedure and why she had to travel to Florida to get it. Nonetheless they seemed impressed, and Hannah was proud of herself. An explorer of cutting-edge science felt like a much better identity than a childless boss who found love too late.

Hannah loved being pregnant with these other women. There was so much happiness to share. Two of her fellow coworkers were pregnant with their second child and gave her tips on how to find the best doctors, where to buy maternity clothes, and what to expect in the upcoming months. "You can't live without the Boppy!" one said, referring to a pop- ular body pillow designed to support a pregnant tummy during sleep. Until then Hannah had never heard of the Boppy and felt honored to be part of this new mom's club. Whenever she joked about her age, they responded, "It doesn't matter. You look so young anyway."

Despite all the warnings of complications, Hannah's pregnancy was easy. She had only a few minor bouts of morning sickness. And an early blood test and ultrasounds to measure the spine and amount of fluid behind the baby's neck—which in large amounts can be a marker of Down syndrome and other chromosomal abnormalities—showed the baby's development was normal, which meant she could skip the inva- sive amniocentesis test. She was ecstatic when she hit the three-month mark and was out of the danger zone for miscarriage. Her doctor warned her that women with a history of anxiety disorders were vulnerable to a relapse during the last trimester. Just as she had promised Nate, Hannah researched the name of a specialist who dealt with such disorders as well as postpartum depression.

Her pregnancy was a source of joy, but she couldn't quite come to terms with the fact that she would be bringing home a living, breathing baby in six short months. It finally became real during her sixteen-week ultrasound, which was scheduled the same day she turned forty-five. Her eyes filled with tears when she saw the translucent figure swimming on the black screen and watched the technician measure the baby's head, neck, heart, spine, arms, and legs: all perfect. She showed Hannah the baby's ten fingers and toes. She checked the fluid at the back of the neck again. Still no indication of Down. Then she aimed for the baby's pelvic region. Hannah held her breath as she waited for the woman to jigger the ultrasound probe for several more minutes. "If it was a boy, you'd see a bump here," she finally said. "There's no bump here, so you're having a girl."

"Oh my God," Hannah exclaimed as tears flowed freely. "I'm having a girl!" She would have been grateful for a son, but she had really wanted a daughter. She wanted to be as close to her daughter as she was with her mom, who was her best friend and confidante. They chatted several times a day about everything from current events and work gossip to cooking tips. Hannah knew how to have intimate relationships with women; she didn't feel as confident being a mother to a boy. They never seemed to stop moving, and her brother definitely didn't talk to their mom as much as she did.

Hannah's mom, who had attended the ultrasound, hugged her daughter tightly, and they both squealed and sobbed.

Nate was quiet. When the commotion died down, he asked Hannah what he usually did during big moments, "Are you happy?"

"I'm so happy! I can't believe it's a girl," she responded animatedly in an attempt to engage him.

"Well, as long as the baby is healthy, we're in good shape," he said, trying to sound upbeat. Hannah knew he had wanted another boy, but she suspected the real reason for his disappointment: he had hoped to share that moment privately with her. "I want it to be just you and me," he had said when she asked if he minded if her mom tagged along to the appointment. "But you can bring her, if it's important."

Hannah now questioned whether it was a good idea to let her come. She should have anticipated that her reserved husband wouldn't be comfortable showing strong emotions with her mom standing there. She already doubted how much Nate wanted this baby. She trusted he would be a devoted father and adore another little girl. She loved looking at old photos of him with Holly as a baby, feeding her a bottle or sleeping with her nestled on his chest. But the distance between them felt greater than ever.

While he rushed back to work, Hannah and her mom went out to lunch to celebrate.

Hannah's friends tried to encourage her, arguing that men often act detached because they resist change or are worried about money, but they always rally in the end. *Their* husbands did, they said. She knew Nate was under pressure to figure out how a blended family would function. He had to pay attention to her and the baby and still fulfill his responsibilities to his older kids. Not only were there logistical challenges, such as juggling Jason's sports practice schedules, but he had to convince Holly she wouldn't lose her place as his daughter. The sixteen-year-old was not happy about the arrival of another girl in the family.

Hannah's mom piled on. "You didn't see the look on his face during the ultrasound," she said. "He's so happy and so in love with you."

Why couldn't Hannah see how much he loved her?

Monica

With Monica's endometriosis at bay, she changed her Match status from "not sure" about having children back to "someday" wanting them. She dutifully continued to go on several dates a month, but she didn't feel any chemistry with anyone. She was tired of sitting across from men drinking the same glass of Merlot and having the same conversation about her work, travel, and hobbies. She no longer got that nervous little feeling in her stomach before dates. She didn't have that same sense of wonder if he could be "the one." And she didn't want to go out for long dinners or engage in make-out sessions to find out. Her heart hadn't hardened; it had just learned to open and shut with remarkable efficiency.

What made her happy was spending more time with her young nephew and niece, who called her "Little Mom-Mom." She toted her niece around on errands and was amazed at how much the little girl looked like her at that age. Monica felt a burst of pride when people assumed she was her daughter. She also enjoyed going to her nephew's soccer games. When she had imagined attending her own children's soccer games in the past, she wondered if she would resent how they took up large chunks of her Saturdays. But she loved watching him run up and down the field, his chubby little legs swallowed by shorts and shin guards. She cheered with all the other mothers who seemed to be about her age. Although she rarely felt the discomfort that many childless women experience, of being out of sync with their peer group, especially in the suburbs, she was reassured nonetheless.

Once again Monica felt the pull to have a baby. Once again her sister urged her to go ahead and have one. "Mon, what are you waiting for?" she asked. Her sister, who was married with two teenagers, had joked over the years that Monica should get pregnant using donor sperm, to which Monica had always replied, "You're crazy!" Her former therapist, Roni, had also asked her a few years earlier if she would ever have a baby by herself. At the time, Monica had wanted a baby with Adam so badly she couldn't imagine being a single mom. She thought a baby would make them a family.

But now that she wasn't in love with anyone, the idea of trying to quickly connect with one of these Match men in order to have a baby had started to feel more and more absurd. She had the rest of her life to find a man. But at age forty-two, she had only a short time left to have children, especially since her endometriosis could return within a couple of years. Her illness had in effect stolen the extra time that egg freezing had given her. Her illness also reminded her how vulnerable her body was. If it came back, how would she have the energy to run after small children?

Monica still wanted a man in her life, but she began to wonder if it made more sense to have a baby first, in case one didn't come along in time. Her future husband could still love her baby. They could still be a family. She thought of her college friend who had seven-year-old twins and was seriously dating a fifty-year-old man with grown kids who welcomed being a stepfather.

The idea of using donor sperm had always seemed like a distant last resort full of complications and conditions. But it slowly dawned on Monica that she could visit a sperm bank, thaw her eggs, and become pregnant right now. She could have an actual baby in her house by this time next year.

When she received her annual egg-freezing storage bill that December, she wondered how many more times she was going to mail in a check for several hundred dollars. What *was* she waiting for? On New Year's Eve she vowed to make a decision during the next year. In the meantime, she continued dating. A few months later, she met two promising men. She met the first one, a muscled, never-married fifty-year-old, at a bar while she was out with a group of friends. She felt intensely attracted to him and loved his easy sense of humor. But on the third date, she forced herself to stay on message, saying, "I'm nearly forty-three. When I date now, I have to know that having kids with you is an option."

"Can I get back to you on that?" he responded. Monica had a sinking feeling.

A couple of weeks later, she pressed him again. "I need to know," she said during dinner. "I don't want to waste your time or mine."

"My priority is to meet someone," he admitted. "Children are not my priority."

The following week, at a happy hour at a local steakhouse, she met a cute forty-three-year-old man who also had never been married. This time Monica brought up the subject on the second date. "What are you looking for?" she asked. "I'm not into dating just to date. I'm looking for a serious relationship."

She was relieved when he said he was ready to settle down and have kids. But she quickly soured on him when he accused her of being out with another man when she didn't respond to his text one evening. She left the date defeated. Even if they both wanted the same things, there were so many other factors that could wreck romance. When Monica was younger, she assumed that most guys she met wanted kids. Now she mused how ridiculous it was that at her age simply wanting kids should be a sufficient reason to seriously consider someone as a life partner.

"That's it. I'm done," she said aloud to herself as she walked to her car. She was used to the disappointment of not feeling attracted to someone she had exchanged a few emails with over the Internet. What bothered her about these last two prospects is that she had met them in the real world and had felt a rare chemistry. They *both* seemed promising, and they *both* turned out to be very wrong for her.

A few weeks later, Monica turned forty-three. As usual, she made a list of her accomplishments from the past year and a list of her future goals, which included a husband and a baby. If she started the pregnancy process right away, she would be forty-four by the time she gave birth. Even if she met the right man tomorrow, she wouldn't be on track to have a baby for at least a couple of years, and she would be inching that much closer to fifty. For the past decade, she had been aware that she didn't have time to waste, but it seemed as if she'd always had a few years to play with. Now she was overcome with a conviction that if she didn't act—if she continued to hope and wait for the perfect scenario—she would miss out on everything.

This certainty even overrode any doubts she had about being a single mom. Monica hadn't yet thought through the details of how she would juggle a child and a job. A friend had told her, "Mon, you can do it. You don't need a man. The women do all the work anyway." She didn't yet know how she would find time to date or find her baby a father figure, but she hoped to benefit from a new dating pool: divorced dads. Although she had been open to dating dads all along, she wasn't particularly enthusiastic about them. She didn't like being fit into their child custody schedules or following them around to their kids' sports activities on weekends. And she sensed they thought she was inflexible and frustrated. She guessed they usually ended up with other single parents who better understood their lifestyle.

Maybe she and a single dad could blend their families. Maybe she would meet an older guy who had never had children but wouldn't mind being a stepdad, since he wouldn't have to deal with caring for an infant or financially supporting a child. Or maybe she would find an older man whose children had grown and missed having a young one around. Whatever the case, her child *would* have a father—just not a biological father and just not right away. She also hadn't imagined how she would tell her child about the sperm donor, or explain to people why there wasn't a man in the picture. They were just details, and she would figure them out.

Monica decided to have a baby.

The only thing she wasn't sure of: how to tell her mother.

Her parents had separated since she asked their opinion of egg freezing in a Chinese restaurant so many years ago, so she met them individually. She started with her dad first. "Dad, I have something to tell you," she announced at the family's shore house one weekend. "I'm going to have a baby using my frozen eggs."

"That's great," he exclaimed. "I just have one question. Can you make it a boy?"

Later that weekend, she overheard him asking her sister if she had picked out a donor.

Monica planned to tell her mother the following week, after an outing with her sister to Manhattan to see the Broadway musical *Memphis*. She knew that her mother would support her, but she was still nervous. Monica wasn't afraid of disappointing her; she just didn't think her mom had imagined that her beautiful younger daughter would be choosing the father of her grandchild by trolling a database for sperm. This wasn't the life she had worked hard to give her daughter.

When they retuned to Monica's house after the performance, her mom said, "I hate to spoil the good time we've had, but you and your sister need to know what's happening with me and your dad."

She told them they had set up an estate and will. "Monica, you're a special case," she said gently.

"Whoa!" Monica shot back. "Did you write something in there for me and my future kids?"

"Don't worry," she replied. "We put something in there if you have kids."

As Monica imagined an asterisk next to her name on the will, she became even more confident of her decision. She was grateful that her parents would leave her an inheritance, but the discussion made it even more clear how alone she was in the world. Her sister and brother had spouses and children to take care of them when their parents were gone. She had to think about who would take care of her when she was old.

She changed the conversation. "Well, Mom, I need to share something with you," she said. She looked at her sister, who was staring at the floor. "Did you already tell her?" Monica pressed.

Her sister nodded.

Relieved, Monica turned back to her mom and continued, "Well, I don't want to be fifty years old and look back and kick myself."

Monica waited for her mom to grill her. Instead her mother said, "I understand. Don't worry. I have plans to retire next year. I'll help you."

Monica was dumbfounded as she relished the moment. Her mother wasn't questioning her about money or dating or schedules or the success rates of frozen eggs or contingency plans. She had simply said, "I'll help you."

Monica thanked her and asked if she'd also help her narrow down her sperm choices. That evening she felt a surge of love for her mother as she realized that her mom had complete faith in her. "I know you always land on your feet," she said as she kissed her daughter goodbye.

Monica had hoped picking a sperm donor would be as fun as browsing through Match profiles, but she was quickly overwhelmed with information. The database for California Cryobank let users search by everything from donors' eye color to jaw shape, grade point average, and personality type. She could see their baby pictures and read their personal essays. She could even choose a donor who looked like the celebrities Bradley Cooper, Jon Hamm, or Ben Affleck—even Jon Gosselin and Bill Gates. But she wasn't interested in pretending she was having the baby of a fantasy boyfriend. She had to be practical about what traits would give her child the best advantages in life. She thought that height was especially important for a son, and at five-two, she wanted a tall man to balance out her short Filipino genes. She had always dated white men, so she selected a donor of European descent who was at least six feet tall as part of her search criteria. Then there was health. Each profile came with four pages of detailed information about every kind of disease, condition, and addiction on both sides of a donor's family, including grandparents, aunts, uncles, and cousins. Monica remarked that if she was in love with someone, she would have overlooked his health history and hoped for the best. But she loved being able to screen out mental illness and alcoholism, which she worried could be passed down to offspring. She spent many evenings poring over profiles and making checklists. The whole process appealed to her desire for control and love of efficiency. Maybe it was better to have the baby she *wanted* first and find love second.

Eventually she narrowed down her choices to three college students with stellar family medical histories: a mechanical engineer, a chemical engineer, and a literature major. "Aren't you worried they have no work experience?" her mom asked. Monica thought it was weirder that she was choosing among guys who were two decades younger than she. But she reminded herself that she wasn't dating them; she was just buying their genes. She was most drawn to a twenty-four-year-old chemical engineer who was six-one, weighed 190 pounds, loved to hike, and had an artistic side, and who included photos of Swiss mountains and a sketch of a man and woman kissing with the word *love* written underneath. She listened to a recorded interview and loved his deep voice and thoughtful answers about how he had been inspired by his uncle to be an engineer and wanted a family of his own in about ten years. She thought this was important information to tell her future child about the donor dad. Plus, his baby pictures were adorable. She ordered three vials of his manhood.

When Monica was younger, her baby fantasies consisted mostly of walking with her husband, who was pushing a pram in a Copenhagen park or hiking with a baby carrier strapped to his chest. She used to feel stricken with longing when she saw beautiful families in church. But she didn't feel pangs of loss anymore. She thought about strolling through her town's picturesque village square while carrying her baby and walking Milo, a German shepherd mix she had adopted a year earlier. For now, they would be their own little family.

In November 2011 Monica's heart pounded as she stood before the mailbox, looking at the envelope containing the forms she had signed and gotten notarized, authorizing the release of her eggs from the storage facility to her new fertility clinic, Reproductive Medicine Associates of New Jersey. (She switched from IVF New Jersey after learning the center was no longer affiliated with Extend Fertility and that none of their egg-freezing patients had come back to thaw their eggs. She liked that doctors in the RMA network had some success stories by then.) She held the documents for nearly a minute before she dropped them in the mailbox.

Kelly

For the past four years, whenever Kelly attended baby showers or admired moms pushing babies in shopping carts, she had quieted her longings by saying to herself, "That's going to be me someday. I have my eggs."

But her eggs were gone now. She no longer had a backup, an ace in the hole, a secret supply of youth that meant the laws of aging didn't apply to her. She was suddenly a forty-three-year-old woman with the same options as everyone else.

She didn't want to waste any time grieving. She had to find out if it was too late to try IVF. She knew the success rates for fresh eggs were dismal for her age, but she purposefully didn't look them up. Many women undergoing IVF are obsessed with stats, but Kelly often found comfort in not knowing. Every woman was different, she told herself, and she knew trying to make sense of all those numbers could drive her crazy. Besides, those statistics were based on women who had been diagnosed as infertile, and Kelly had been off the charts with nineteen eggs at age thirty-nine. Maybe she was still a super-producer and had a better shot than her peers. Her mother-in-law joked that fresh was always better than frozen anyway, as though she was buying fruit in the supermarket.

Kelly liked Dr. Winslow, but she and Dan wanted to find a clinic closer to their home and selected one in Atlanta called Reproductive Biology Associates, which was only three and a half hours away instead of the seven hours it took to drive to Jacksonville. The clinic also had a large donor egg program, in case they had to resort to that option. Kelly was nervous about whether the clinic would accept her as a patient and was heartened when Dr. Carlene Elsner told her she could start immediately, as long as her hormone levels were still in the desirable range.

Kelly had her blood drawn on the third day of her next period, just as she had done at Winslow's office. Whereas her FSH level had been 8.9 when she was thirty-nine, it was now up to 11.5. The ideal was below 10. She was considered borderline, though the nurse told her only levels above 20 would be discouraging.

"I'll be honest," the doctor told her. "Your chances of success aren't great. And if you want to use your own eggs, time is of the essence."

Kelly would have better odds of bringing home a baby at her age if they tried donor eggs right away, but she wanted one more chance at conceiving a biological child and was grateful Dan was willing to pay $15,000 to try IVF. She marveled at the thought of creating a baby who behaved and looked like her and her husband. She enjoyed noticing similarities between Dan and the twins and wondered what mix of characteristics their baby would have. His sparkling eyes? Her vibrant smile? She hoped their child would share their love of antiques.

Kelly didn't feel forty-three and hated it when people said, "Having kids at your age? That would be tough." In fact if this IVF round worked, she would be forty-four by the time the baby was born, and Dan would be forty-nine. She had waited so long to have kids that she looked forward to feeling exhausted at the end of a busy day with her little ones. If motherhood might have been easier in her thirties, she wouldn't have known the difference.

During her fresh IVF round, Kelly *was* still an egg super-producer, that is, for a forty-three-year-old. She had six eggs harvested, five of which grew into embryos; the doctor transferred all five into her uterus and told her that she had only a 5 to 10 percent chance of one resulting in a live birth.

When Kelly learned two weeks later that the attempt was a failure, she wasn't surprised, but she was disappointed. Even though the success rates were low, *some* women still got pregnant. Why couldn't she have been one of them? This time, though, she didn't cry. She just tried to figure out what to do next.

Kelly and Dan had talked about exploring donor eggs, which had a 60 to 70 percent chance of success. They liked those odds but wondered if they should give IVF one more try since Kelly still had quality eggs. This was their very last chance for a biological child, and they wanted to make sure they had given it their best shot. Dan said he would look into payment plans offered by medical financing firms, and they agreed to use the money from the sale of Kelly's house to pay off the loan. It helped that their health insurance covered the bulk of the drugs.

Kelly was so relieved they were on the same page financially. They weren't going to let the fear of debt stop them from pursuing their dreams, even if they joked about future lean years of peanut butter sandwiches. On the other hand, they wouldn't lose their home chasing every promise of success, no matter how low the odds. She felt their plan was prudent: if the next round of IVF didn't work, they would try two rounds of donor eggs. Before they resorted to adoption, Dan estimated they would have spent nearly $100,000 trying to have one baby.

Kelly wondered if they should start the foreign adoption process, which could take years, just in case. She worried that American mothers looking to place their babies would choose a younger couple, but even overseas adoption didn't seem as easy as she'd always thought it would be, considering that each country had its own restrictions. For example, some countries capped the age of prospective adopters at forty-five or required there be no more than forty or forty-five years difference between the age of the parent and the age of the child. Kelly had always admired a little adopted Chinese girl she saw at church. Although she and Dan met China's age requirement that both adoptive parents be under fifty, their marital histories might make them ineligible. A new law specified that if either member of a couple had been previously divorced, they had to wait until their fifth wedding anniversary before applying. Kelly would make the deadline at age forty-eight, but Dan would be fifty-three.

In the meantime, Kelly searched the Internet for tips to improve her chances of making this last round work. Some advice, such as taking certain herbs or eating fewer carbs, seemed dubious. But she was curious when she read on an online message board about a woman who had visited a New York City doctor who claimed he could significantly increase the odds of a successful IVF pregnancy. He didn't focus on a woman's age; he focused on her bacteria. Dr. Attila Toth (not related to Dr. Thomas Toth of Massachusetts General Hospital) is a Hungarian urologist who specializes in "unexplained infertility," a diagnosis given to a third of IVF cases in which the cause of infertility is unknown. Dr. Toth believes that inactive strains of bacteria, especially Chlamydia, exist undetected in many women and can prevent the implantation of embryos. He doesn't think it's a coincidence that the rates of sexually transmitted diseases and infertility have risen in tandem. Plus, the fact that women are marrying and having children later means they've had more opportunity to collect multiple strains of bacteria.

Toth's treatment is extreme: a ten-day cleansing regimen consisting of antibiotics injected into the uterus for an hour a day. The patient also wears a pump that shoots another cocktail into her bloodstream twenty-four hours a day. For maximum results, he advises her partner to undergo a similar cycle of antibiotics injected through his rectum into his prostate. Toth's success rates are impressive: in his unpublished study of sixty-three couples who both underwent treatment, twelve became pregnant naturally, and seventeen delivered babies through IVF.

Kelly wasn't sure what to make of Toth. Indeed it was hard to understand why a purported medical breakthrough wasn't standard practice at fertility clinics around the world. But she was mesmerized by the stories she read on his Yahoo! group. She couldn't help but be moved by the woman in her forties who had five failed IVF attempts but conceived on her own after treatment with Toth. When she learned about three other women who gave birth to IVF babies after going through Toth's program, Kelly scheduled a phone consultation.

She knew she was being persuaded by success stories. But she was in a bind, and this man was offering a solution, in the same way Winslow had offered her one four years earlier. This was her last chance *ever* to have the biological child she had dreamed of her entire life. His treatment was just $4,000 more.

A couple months later, after Dr. Toth tested her for Chlamydia, Kelly was in Manhattan with an IV pump hooked up to her arm and wearing a fanny pack of meds. Every morning, she visited his office and grimaced while he threaded a catheter through the pinpoint opening of her cervix. She then lay on the table for an hour reading gossip magazines while antibiotics coursed through her. Back home, she and Dan took several cycles of oral antibiotics. Dan was spared the prostate injections since they wouldn't be trying to get pregnant naturally, but he still had to use a condom during sex for three months to prevent the exchange of any more bacteria.

Kelly was satisfied that she was doing everything possible to give herself the best chance of having a baby, but she was physically exhausted. One antibiotic made her sick with flu-like symptoms, and she lost five pounds. Another left a copper taste in her mouth. Her Atlanta doctor found heavy growth of *E. coli* in her uterus and prescribed two more medications. She was also weary from driving seven hours roundtrip to Atlanta for doctor's appointments. One afternoon she surprised herself when she burst into tears after having trouble scheduling another test to look for defects in the lining of her uterus. This was a different test than the one she had undergone at Winslow's office and was usually recommended for patients who had experienced many failed pregnancies. Nevertheless, the fact that she was scheduling this test *now* infuriated her. Why hadn't her doctors suggested this test before she had used up all her frozen eggs and spent all that money?

It was unusual for her to feel so edgy and tearful, and she tried to warn Dan about her moods. "I don't know what's wrong with me," she'd say. "I feel like I could cry for no reason." She wanted to make an effort to share her feelings as openly as possible with him since she was aware that infertility could be stressful on a relationship. They had been married for just a year and a half and had already endured so much.

Dan liked to handle crises by taking charge, so Kelly let him schedule her appointments and fill her prescriptions. Sometimes when she arrived home from a doctor's visit feeling down, he lit candles on the side porch and served her dinner. She wanted him to feel he was part of the baby-making process, so she asked him to administer her hormone shots and keep track of her medication schedule. He commented that his sperm aspirations made him feel more dedicated than other men who contributed their genetic material into a cup in a clinic closet. Still, he confessed that he felt bad about his vasectomy. They could have been trying naturally all this time.

During the first week of December, Kelly produced ten eggs. Seven fertilized, and all were transferred into her pristine uterus.

Sarah

I have always been comforted by numbers. I applied to fourteen colleges during my senior year in high school and was on six Internet dating sites at one point. Volume was a form of protection because it seemed as if you always had another chance. I wanted two children and would need thirty eggs for each, if I followed Dr. Tan's advice. I had twenty eggs already, and I had two cycles left in Montreal. I knew my last cycle's yield of fourteen eggs was unusual compared to the six from the first round, so I conservatively estimated that I might net ten more eggs per cycle. That would bring me up to forty. Not enough.

The truth was I didn't know if there was ever a number that would make me feel secure. Even if Paul agreed to try for a baby right away, I didn't know if I would want to waste valuable freezing time. Pregnancy would tie up my body for nine months, and I had one year left to retrieve the best eggs possible.

I researched doing another cycle at my New York clinic, Reproductive Medicine Associates, and learned that Extend Fertility charged the discounted price of $9,000 for subsequent rounds. I wondered if I should just buy another round in Montreal—in addition to the two I had left—but a single round cost $6,000, and considering the extra money for monitoring, flights, and a hotel, I wouldn't be saving very much. Besides, I didn't want to go to Canada in the dead of winter, and RMA's head embryologist was claiming extraordinary egg survival rates with his slow-freezing method. I asked Dr. Mukherjee to follow Dr. Tan's drug protocol and was overjoyed when I produced sixteen more eggs. That put me at thirty-six, enough for one and one-fifth kids.

In Bologna I had met Dr. Masashige Kuwayama, a Japanese fertility doctor who had become a bit of a celebrity in the egg-freezing world for his high egg survival rates with vitrification. When he learned that I lived in New York, he gave me the card of his American partner, Dr. John Zhang, who had an office on Park Avenue—a detail he repeated several times. I didn't need yet another doctor, but I looked him up anyway. He had a couple dozen births from frozen donor eggs under his belt. He also had been on the front page of the *Wall Street Journal* in 2006 because of his unorthodox approach to IVF, called "minimal stimulation." Instead of injecting expensive hormones called gonadotropins for ten days, patients endure just a couple shots supplemented by the significantly cheaper ovulation-induction pills clomiphene citrate, also known as Clomid. The method has been used widely in Europe, but American doctors have been reluctant to embrace it. The reason: patients produce fewer eggs. However, doctors like Zhang believe the eggs they do produce are the ones most likely to turn into babies. Here's how it works: At the beginning of each menstrual cycle, a woman's body releases a dozen or so immature eggs as candidates for ovulation. But the body chooses only one to send to the fallopian tubes to be fertilized (the baby-maker egg, so called because it's the egg given the chance to make a baby). According to Zhang, each batch contains the baby-maker egg and a couple of other good eggs that are likely to develop into viable embryos. He aims to coax out these lucky few eggs rather than blast out the entire batch, as is done in traditional IVF stimulation. He says there's no point in harvesting the other low-quality eggs, especially at such a physical and financial cost to patients.

Despite this logic, minimal stimulation is believed to have lower success rates than traditional IVF. But it's hard to compare them. For example, Zhang says his rates are lower than those of traditional clinics because he transfers fewer embryos at a time to reduce the chances of multiple births. He also accepts older patients and those with undesirable FSH levels who might not make the cut at clinics who carefully select their patients to keep their success rates high.

It didn't make sense to compare the success rates from single rounds of traditional and minimal stimulation anyway. That's because Zhang's fee for his "future mother" egg-freezing program was $8,250 for three rounds. He explained that I would produce about the same amount of eggs over his three minicycles as I would in one traditional cycle. Plus, these eggs would cumulatively be of better quality, and Clomid would be easier on my body than hard-core hormones. Those drugs cost about $500, compared with several thousand for injectable gonadotropins.

I told my mom about his approach and asked if she thought I should freeze more. I already had thirty-six eggs. If I netted twenty more from my remaining two rounds, that would bring me up to fifty-six. "The Montreal doctor said it was ideal to have thirty eggs saved per child," I told her.

"Go ahead and do it," she said with a sigh. "We'll cover it."

As much as I tried to persuade myself that I was acting prudently, I knew I was being obsessive. Instead of calming my anxiety, the quest for more eggs left me feeling frantic. It became a reaction to every lurch in my gut. Scared of losing Paul? Freeze more eggs. Scared of losing Claire? Freeze more eggs.

Besides, freezing felt good. With every egg I stashed away, I moved further away from the persona of the sad girl whose baby-making window was closing and back to the hopeful girl who always had *someday*.

Dr. Zhang's minimal stimulation *was* slightly easier on my body, although one month Clomid turned me into such a foul human being that I unleashed on Paul for choosing the wrong corner from which to hail a taxi. During the three cycles I produced eight, four, and three eggs. I was now up to fifty-one, with two more rounds to go in Montreal.

My friend Janelle had traveled with me to Montreal during my first cycle; Paul had accompanied me on my second trip and had offered to return. But I was thankful that my mother came with me on my last trip, when I needed her most. She held my hand during surgery and brushed the hair away from my eyes in the recovery room. "This was the right thing to do," she said. Later that evening, in the kitchenette of our bed-and-breakfast, she made us dinner from food we had collected at the farmers' market: sausages, squash, and exquisite September tomatoes. We ate while shouting out the answers to *Who Wants to Be a Millionaire?* In the middle of the night, I watched her sleep with her book on her chest and the light on, and I felt a surge of love for the person who had insisted on flying cross-country to take care of me. I was also grateful that, having given birth to me at age twenty, she was only fifty-eight now. I had the best of both worlds: a young mother and the chance to have children in my forties. My own children would not be as lucky.

I had acted as if my mother harangued me about settling down and popping out grandchildren, but I sometimes forgot that she had dreams for me too. For the past decade, she had kept a folder of wedding ideas. She had saved furniture and dishes for me. She had painstakingly washed, mended, and ironed my old baby clothes for my future children. I knew she wanted grandchildren who looked like me. But she also just wanted me safe, securely entrenched in the social order, with a husband and children to look out for me when she no longer could, not floating around unattached in New York City, dating "maybe" men and freezing my eggs. I would want the same for my children.

She speculated whether I would be able to sell my extra eggs once I had completed my family. "Mom, no one's going to want my old eggs," I said, laughing at her entrepreneurial spirit, as if one could sell unfertilized eggs like concert tickets. "Women want egg donors under thirty-two."

"Yes, but these are *your* eggs," she insisted, as if women would discount the age factor if they only knew how many speech and debate tournaments I had won in high school. I was touched that she thought my eggs were special.

I finished my last two freezing cycles in Montreal—sufficiently doped up—a safe several months before my deadline of my thirty-ninth birthday. I had netted eleven and eight and now had seventy eggs frozen by two different techniques, stored in three different places. I was done. I had enough eggs for seven to twelve tries, depending on how many eggs my doctors decided to thaw per attempt. Only time would tell whether my stockpiling was sheer lunacy or good planning. In the meantime, I cringed to think of the money—more than $50,000—that could have covered weddings, down payments, and school loans. I had spent all of my savings, and I'm extremely fortunate to have parents who could help me. Egg freezing en masse isn't an option for many women.

My body also paid a dear price. My legs and hips were bruised from the generic hormone injections I had sourced through a British mail-order pharmacy and that, unlike my other drugs, had to be injected deep into my muscles. (The pain was worth saving thousands of dollars.) I had gained about eight pounds from three rounds of Clomid with Dr. Zhang. My hair was thinning, a side effect from one of the hormones. My cheeks had dark patches from a condition called melasma, the result of eight cycles on hormones. On top of it all, a head cold caused me to sneeze frequently, which made my tender ovaries spasm. I was sick of the bloat, which started on about the fifth day of stimulation and lasted until my period came nearly two weeks later. I estimated that I had spent approximately one-third of the previous year wearing fat pants.

I wondered if my ovaries were scarred from being punctured so many times. And I couldn't imagine the toll of the hormones on my body. Several fertility doctors told me there was no limit to the amount of hormone-stimulation cycles a body could endure, but one study of women who had used Clomid in the 1970s found a link between the ovulation drug and ovarian cancer. One Danish study, however, was reassuring; researchers tracked more than fifty thousand women who underwent fertility treatments for fifteen years and found they faced no more risk of breast cancer than women in the general population. In any case, my ob-gyn recommended taking birth control pills, which are thought to decrease the risk of reproductive cancers, as a countermeasure. Interestingly, she said that giving birth was another form of protection, as women who never deliver children are at greater risk for such cancers.

My body would eventually heal. For now, I felt triumphant. I had safeguarded my ability to have children, and I was free to be back in my bubble with Paul.

7. *Reckoning*

Kelly

A few days before Christmas, Kelly had her blood drawn to test whether her final IVF attempt had succeeded. She learned the results as she drove the twins' church youth group around town to deliver cookies and sing Christmas carols.

This was her fifth time receiving such a call, and she was used to bad news. She didn't even pull over to take the call from her Atlanta fertility clinic out of earshot of the kids.

But this time the nurse said, "Congratulations, Kelly! You have a positive pregnancy."

Kelly was speechless. At nearly forty-four, she had never heard those words before, and they sounded unbelievably sweet. She dropped the kids at the next stop and stayed behind to call Dan, who shouted over and over, "You're kidding me! Are you sure? I can't believe it!"

She drove the youth group back to church, then headed to Dan's office. He ran out into the parking lot and jumped into her SUV. They clung to each other and sobbed. After all this time and heartbreak, they were finally going to have a baby.

Kelly had the familiar pinching feeling in her uterus she had experienced after her other transfers, but this time the sensations lasted longer and were more intense. She thought that was a good sign but was concerned she didn't have any of the other symptoms, such as nausea, fatigue, tender breasts, headache, or backaches, that are common in early pregnancy. She knew the miscarriage rate was high for her age, so she embraced a surreal state of cautious excitement and decided to wait until she was further along before she shared the news with anyone. Privately she gave Dan a silver baby cup.

Every morning she woke up and wondered if she was still pregnant. The ultrasounds told her she was, but she still had a hard time accepting the fact, even when she saw the sac on the screen that the technician told her contained an embryo the size of a grain of rice. When she finally heard the whooshing sound of the heartbeat, she was overcome with emotion as Dan squeezed her hand and kissed her forehead. The news sank in when she was told to make her next appointment with her ob-gyn in Charlotte. There was no reason for her to return to Reproductive Biology Associates.

Her doctor told her that if she could make it to nine weeks, her risk of miscarriage would drop to 5 percent. At eleven weeks, it would be less than 2 percent. Then she could announce the news. Kelly was most excited about telling the twins and had bought a card game that suggested fun conversation topics to play with kids. She planned to make her own card with the question "Are you having a new brother or sister?" and leave it on their placemats at dinner. She thought about having a party to make the announcement. She couldn't imagine the reaction on people's faces when she told them that, at age forty-four, she was having a baby.

A few days before her nine-week ultrasound, Kelly had an uneasy feeling. She was two months pregnant and still didn't have any pregnancy symptoms. She traveled with Dan to her ultrasound with a heavy heart.

"It's going to be okay," he consoled her.

"I just feel weird," she said. "I don't know if it's nerves or if something's wrong."

It took nearly all her energy to hoist herself onto the examining table, and she could barely feel the technician spread the goo on the ultrasound probe and insert it into her vagina.

"I'm so sorry, but I can't find a heartbeat," the technician said.

Kelly closed her eyes as she listened to the instructions about scheduling an appointment to have the tissue removed from her womb.

Monica

Monica thought it was a good sign that one of the doctors at her new fertility clinic, RMA of New Jersey, was one of the first physicians in the United States to offer egg freezing nearly a decade ago. She asked to be assigned to Dr. Thomas Kim, who had moved to New Jersey from CHA Fertility in Los Angeles a few years earlier. She and Dr. Kim agreed to thaw eleven of her nineteen eggs, which she hoped would give her a few good embryos. If she became pregnant on the first round, she would still have eight eggs left for another child—maybe even with someone she would meet in the meantime.

Monica was surprised at how hands-off the entire baby-making process was. Instead of making seduction plans, she essentially made shipping plans. She arranged for her eggs to be sent from the storage facility in Massachusetts, and she ordered vials of donor sperm over the Internet. Besides taking hormone supplements to prepare the lining of the uterus for implantation, her only other job was to wait by the phone.

When Dr. Kim called, he delivered the news that only six of the eleven eggs had survived thawing, but five had successfully fertilized. He would let the embryos grow for five days until they reached the blastocyst stage, comprising more than a hundred cells, then transfer the best two and freeze the rest. (Doctors are divided on the approach: some believe blastocysts have a better chance of resulting in a live birth, but others think they're more successful with younger patients and that the embryos of older patients shouldn't be allowed to develop in the lab more than two to three days before being transferred.) Monica was thrilled with her windfall of five embryos and started calculating what she would do if she conceived twins. Maybe there would be leftover embryos to freeze.

On the day before they would be transferred to her uterus, Dr. Kim called again. "I'm sorry, but I don't have good news," he said. "The embryos have stopped growing."

"All of them?" Monica asked.

"Yes," he responded and then explained he thought she had a low chance of becoming pregnant from those eggs. Monica had frozen them six years earlier using old technology, which wasn't as good as the current flash-freezing method. She appreciated the context but was puzzled why he was elaborating on it: the new technique couldn't help her now.

She immediately called her mother. "I just talked to Dr. Kim," she reported in a stunned shaky voice. "There will be no transfer. All of them arrested."

"What? What?" her mom responded. "I don't understand."

"Mom," Monica said, trying hard to be patient, "none survived."

"Oh," she replied. "I'm so, so sorry."

Monica reminded her that she had eight frozen eggs left. Plus, Kim said she could still try IVF with her forty-three-year-old eggs. But her mother kept repeating, "I'm just so sorry, honey."

After a few minutes, Monica wanted off the phone. She had to get back to work, and she was uncomfortable with her mom's grave tone. "Thanks, Mom. I gotta go," she said.

Next she called her dad, who sounded more upbeat. "Monica, I will support you. If you want to try again, I will help you," he said.

Monica tried to go about her workday while circulating the arguments in her head that she thought would make her feel better: "You still have more chances" and "Now you have time to find a partner." But she couldn't stop thinking about how devastated her mother's voice had sounded. A few more times after replaying "I'm so, so sorry" in her head, she excused herself to go to the ladies' room. She shut the stall door just in time before hot tears began to flow.

A month later, Dr. Kim's team thawed the rest of Monica's eggs. This time, five of eight eggs survived, and four fertilized. Four days later, three of the embryos had stopped growing. Kim said there was a small chance the last straggler might make it, but it died the following day.

And just like that, after seven years of scheming about her frozen eggs, all of Monica's frozen eggs were gone. What particularly hurt was the fact that she never got a chance to carry them. For months, she had practiced her first act of maternal love by abstaining from alcohol and caffeine and learning new kale recipes. But the little embryos couldn't even make it out of the Petri dish. Her womb, once overrun with endometrial tissue but now perfectly prepared, with a medically thickened lining, had been waiting in vain.

That cold January evening, Monica felt entirely alone. After the condolences from her family and friends dwindled, she longed for a partner to share her grief. She ate a handful of potato chips for dinner, hugged her dog, and sobbed. Even if she had a baby, is this what single motherhood would feel like?

The week after next, Monica was back on eHarmony. She didn't expect to meet a partner to have a baby with before she started her fresh cycle using the rest of her donor sperm, but she figured it couldn't hurt to start the dating process again, in case this one failed. Her father gave her a check for $13,000 and quipped, "If we get a kid out of this, we're going to call him 'Dividend.'" Monica contributed $4,000 for the medications.

Dr. Kim didn't quote Monica a success rate, and she didn't ask for one. Her FSH level was still in the normal range, and she wanted to pursue a cycle, no matter the odds. She wanted to be able to say to herself, and to others, "I tried so hard to have a baby that I was willing to have one on my own. Plus, I even did IVF at forty-three." That list of efforts didn't yet include using donor eggs, which was arguably her best chance of bringing home a baby. As for a biological baby, Monica imagined that she would feel a deep satisfaction knowing she had tried all her options. She would be prudent and undergo just one cycle. If she tried again, it would be with a partner. Maybe then she would be open to using donor eggs. That way the baby would have *his* DNA and another woman's eggs—the opposite of how she was trying to conceive a baby now, using *her* DNA and another man's sperm.

The Society for Assisted Reproductive Technology doesn't even report statistics for women older than forty-two. According to 2010 data, the number of IVF cycles with fresh eggs resulting in live births for women forty-one to forty-two at Monica's clinic was one in five. She would be nearly forty-four by the time she started the cycle. "We can try," Dr. Kim had said. They scheduled her start date for May.

Over the next two months, she responded to a handful of men who contacted her on eHarmony and went on a couple of lukewarm dates. In the spring, she traveled to Hawaii for a work conference. When she updated her Facebook status to "Aloha!" and wrote that she had checked in at the Hilton Hawaiian Village Waikiki Beach Resort in Honolulu, a former romantic interest responded.

"Bar Boy?" she asked herself. "Why is he liking my post? What does *he* want?"

Bar Boy was a thirty-eight-year-old named Doug whom she had met at a local bar at the beginning of the previous winter. Monica liked his "frat boy" good looks and outgoing personality. Plus, he wanted kids. For the following three months, they settled into a routine: every week he met up with her at a bar, and they ended up at his house. At first she didn't expect much from the hookups. She enjoyed the sexual chemistry and welcomed the attention, especially after all her time alone while being treated for endometriosis. But she eventually wanted to know if Doug was a candidate for a real relationship and started pushing him to spend more time with her, even asking outright, "Can we ever hang out when we're sober?" After he ignored several invitations to go to the movies or on a hike, she stopped responding to his texts to come out drinking. When their friends chimed in that he was immature and a waste of her time, Monica wrote him off completely. A few weeks after their last liaison, she ran into him while she was at a bar with her friends and decided to announce to the group that she was planning to have a baby on her own using her frozen eggs that fall. "Wow! I hope it works," he said, adding his voice to the chorus of good wishes. As Monica focused on getting ready to thaw her eggs, she stopped going out. She thought that was the last of Doug.

So she wondered why he had suddenly surfaced again a year later. She frequently received random texts, such as "How are you?" or "We should hang out," from previous dates who wanted to reestablish contact after fading away. She didn't think much of them; they were usually from men who had lost interest in her and now were bored and wanted sex. Or worse, they were newly single after breaking up with the women they had passed her over for. Whatever the case, Monica didn't believe any of these ghosts would suddenly become the love of her life the second time around.

However, she *was* curious about Doug, especially when a mutual friend mentioned that he had recently been asking about her. Monica noticed he had posted a photo of a beach, so she sent him a message. "Where are you?" He wrote back that he was in Florida and asked if she would accompany him to a wedding in a few weeks. She was puzzled but intrigued.

"Sure," she wrote back. "But you know I'm trying to have a baby. I'll have to check my schedule." She was supposed to start her stimulation medications in a few weeks. "Since I haven't seen you in a while, maybe we should hang out. You never did meet my dog."

The following week, they met on a Sunday afternoon for a hike with her dog. Monica loved how affectionate he was with Milo. She enjoyed seeing him again and hugged him good-bye, but she didn't want to go further. She liked to think she was open-minded, but she couldn't be intimate while she was preparing her body to receive embryos made from the sperm of another man, even if the donor's identity was unknown.

A few days later, Doug asked her to go to the movies. As they sat in the dark theater waiting for *The Avengers* to start, Monica broached the subject that surely had been on his mind: "If the IVF fails, I will just move on with my life," she said.

"Well, you know, I can always try to get you pregnant," he replied. "Would it be so bad to be stuck with me for the rest of your life?" Monica didn't think he was serious, but for someone who couldn't go on a real date less than a year earlier, she thought it was a big step that he could at least joke about the subject.

After the movie, she popped the question she thought she'd never ask a man who wasn't the father of her embryos: Would he drive her to her transfer?

He readily agreed and promised to make arrangements to work from home that day. For the next two weeks, he texted her daily, asking her how her day was going and how she was feeling. He even went to her family's shore house for Memorial Day weekend. Monica felt comfortable having sex, and Doug started acting like her boyfriend. She hadn't had a real boyfriend in nearly three years and forgot how much she enjoyed having one. This relationship was different, though; unlike her others, she didn't feel carried away by infatuation. Doug felt like an old friend, or at least one she could occasionally be intimate with during an IVF cycle.

But why now and why him? Monica was confident that men would still be interested in her romantically when she was a single mom. But she assumed that few men would want to be involved with her *while* she was trying to become one. Was it an easy way to try out fatherhood without having to be ultimately responsible for a child? Did the fact that she was undergoing this alone appeal to Doug's caretaker instincts? Did the fact that she truly didn't need him to make her dreams come true suddenly inspire him to want to be needed? Or did he genuinely like her, and was he willing to accept her whole package?

At age forty-three, Monica produced an impressive nine eggs. Six were successfully fertilized, and four developed into embryos. She couldn't believe she was still in the baby game and could barely stand waiting five full days before Dr. Kim told her how many were viable for transfer. Unable to contain her excitement, she searched the Internet for tips on how to improve her fertility. She read about eating the core of a pineapple to improve embryo implantation. She also found a Catholic fertility blessing. On the eve of her transfer day, rational, scientific Monica prayed hard to St. Gerard: "Almighty Creator, hear this fertility prayer and the wishes of my heart. You know my deep desire for a child—a little one to love and to hold, to care for, to cherish. Grant that my body may conceive and give birth to a beautiful, healthy baby in Your holy image."

The next day, Dr. Kim called to say he had canceled her transfer. Three of her four embryos had arrested. They would wait to see if the last one survived, but since the narrow window of time when Monica's uterus was able to receive an embryo would have passed, he said he could freeze it and transfer it later. In the meantime, she had the option of undergoing another round, after which he would transfer all the embryos at once.

A few days later, Monica was shocked to learn that the last straggler had survived. She asked the lab to perform genetic screening to determine if it had any chromosomal abnormalities. The controversial procedure, which involves removing a cell from the embryo for testing, adds several thousand to a patient's IVF bills and can damage the delicate embryo. But some parents prefer learning about potential birth defects before trying to achieve a pregnancy rather than wrestle with the agonizing decision of whether to terminate a pregnancy several months later or hope for the best.

She didn't think much of her slowpoke embryo, which Kim said wasn't even considered top quality, until she received the lab report: the embryo was perfectly normal. And it was a boy.

Monica's eyes started to water. She had always wanted a boy. She couldn't believe she got to learn she might be having a boy before the embryo was even in her body. Suddenly she began to believe she might be having a baby soon.

Monica wanted to try one last round of IVF. Her parents agreed to help her financially, and she produced another nine eggs. Four fertilized. And once again, they all stopped growing. Monica experienced her now familiar cycle of emotions: disbelief, grief, disappointment, and coming to terms with the next option, such as trying her frozen eggs or doing another round of IVF. This time, though, she felt a rare emotion: hope. She had a baby boy waiting for her in the freezer.

Doug promised to drive her to the transfer, which would take place in a couple of months. In the meantime, though, he had started to back off. He returned to his previous behavior of texting her at the last minute to meet up. Sometimes she didn't hear from him for a week at a time. Earlier in her life, Monica might have been hurt; now she was just annoyed. She focused on getting ready to get pregnant with her son.

Sometimes she couldn't shake the feeling that Doug was distant because he was upset about the upcoming transfer. He had made so many joking references about volunteering to be her sperm donor that she wondered if they contained a grain of truth. But she didn't have time to take a break from baby-making to see if she had a future with this man, especially as he was almost forty and still asked her to meet him at a bar late at night. No matter how much she liked the idea of Doug's sticking around, she had to be realistic that at this point all she had was a "friend with benefits" and a reliable ride to her fertility clinic.

On the afternoon of the scheduled transfer, Monica received a call from Dr. Kim's nurse. "The embryo didn't survive the thaw," she said. "Dr. Kim will call you to follow up." Monica went numb. Although she had been told that there is a 10 percent chance frozen embryos won't survive the thawing, she never imagined it would happen to hers. She called her mom, who gasped in shock. When Doug arrived to take her to the appointment, she collapsed in tears, and he quietly held her.

"I've got to get out of here," she told him. He took her to a local bar, and they spent the afternoon drinking fruity martinis. When she asked to be taken home to feed her dog, he offered to spend the night, but she thanked him and said she needed to be alone.

Monica was forty-four now and had vowed to stop pursuing any more IVF rounds, but first she saw two more doctors, just to ask their opinion of her chances. The first said that she had helped only one woman her age get pregnant in the ten-year history of her practice. The second one worried about the effect of her past endometriosis on her uterus. "If I felt I would have success with you," he said, not mincing words, "I'd tell you."

Her quest to have a biological baby was abruptly over. She hadn't anticipated that giving up that dream would sting so badly. She loved seeing herself in her niece. She had wanted to see if her child would be athletic like she was or had the shape of her mouth or eyes. But it was a surreal kind of mourning. It wasn't like grieving a death or a breakup, since she had no memories of the child. It's just that she had always thought that a child was in her future in some way or another, and the finality felt unbearable.

When Doug texted her only once the following week, she couldn't stand his rejection on top of everything else. She wanted clarity in every area of her life and decided she needed to break up with him. She asked to meet him at their favorite bar and delivered the lines she had been rehearsing in her head. "We've been hanging out as friends with benefits for over a year," she said. "But these drunken hookups and silly texting games aren't working for me. At this point in my life, I need to move on. What do you want to do?"

Doug looked stunned, then said defensively, "If you've already decided what you want to do, then why are we talking?"

Monica had spent so much time wondering about his feelings for her that she hadn't seen him for the insecure man he really was. She felt a surge of courage and bravely asked him what she'd wanted to know all along: "Do you want to give us a chance? You know, be exclusive?"

His face instantly softened, and they finally had a real talk. Over the next few hours, Doug revealed that he had been interested in her from the very beginning but had been turned off when she announced her intention to thaw her eggs. He was surprised by how much he had missed her and thought he could handle her trying to have a baby with donor sperm. But he had to be honest: it did bother him. "This baby thing is screwing with my head," he confessed.

That night they decided to become a real couple. Thereafter they slept nearly every night at each other's houses, watched favorite TV shows, went to Home Depot, and planned a winter vacation to Jamaica. Doug grilled her steaks for dinner, cooked her eggs and bacon for breakfast, and, just as he had hinted all along, stopped using condoms and tried to get her pregnant. "You'll probably be peeing on a stick soon," he promised.

Monica was touched at his optimism, that he really believed he could overcome her forty-four-year-old biology. She didn't have the heart to tell him it was nearly impossible. For once in a very long time, she stopped thinking about babies and enjoyed making love to the man she was falling for. One night, as she drifted off to sleep, she cuddled up to him, felt the warm body of her dog at her feet, and thought, "This is enough for now."

Hannah

Hannah entered her third trimester with no health complications or depression, and she finally began imagining what it would be like to be a mom. She envisioned carrying her infant daughter along the shores of Lake Washington and walking her kindergartner to the elementary school near their house. She opened her hope chest and found the green yarn and baby cardigan pattern she had brought back from France years ago and asked her mom to start knitting. She took out the matching felt shoes.

She and Nate decided to name the baby Sophia Grace. Hannah started talking to her on her drives to and from work. "How's it going, Soph?" she'd say. "How are you doing in there?" And she started talking about Sophia in front of Holly and Jason. More important, she referred to the role Sophia would play in their lives, calling her "your little sister." Hannah wanted to make the point that they were going to become a new family; she didn't want them to view her and the baby as their father's "second family." She hoped Sophia would be the glue that brought them all together.

Holly's curiosity started to replace her resistance, and she asked Hannah lots of questions about what it was like to be pregnant: How does it feel to have another person inside you? What do kicks feel like? Why do you eat so much for dinner? Hannah told her she loved being pregnant and was astonished by how quickly her breasts and tummy had grown. She tried to articulate how amazing it was to create a child in her body—never mind one conceived from an egg frozen nearly seven years earlier. Two months before Hannah's August due date, Holly asked Hannah if she would look over her college application essay. Hannah welled up when she read that Holly was looking forward to having a sister.

Hannah often thought about her remaining thirteen eggs in Jacksonville and felt protective of them. She didn't know whether she and Nate would try for a second child. Even if they didn't, she didn't want to dispose of them. Her mom often reminded her that if she hadn't frozen them, those eggs would have disappeared during her menstrual cycle. But for Hannah, they seemed symbolic of how she had advocated for herself. Besides, what if someone in her family wanted to use them? She resolved to pay the $200 yearly storage fees indefinitely.

Nate loved Hannah's pregnant body and often photographed her swelling belly. When they cuddled at night, he gave her backrubs or rested his hand on her belly so he could feel Sophia's kicks. Hannah melted into his arms and felt as if she and their baby were swaddled in love. The feeling was as good as she had always imagined it would be.

Why, then, was she unable to shake her anxiety that Nate had agreed to have a baby only for her sake? When she looked at the evidence, he *was* coming around, just as everyone had promised. He installed a bench with drawers in their room so she would have a place for the baby's clothes. (Sophia would take over Holly's room after she left for college.) Even though he originally balked at joining her at a labor class, saying, "I'll go, but I already know what it's all about," he went and asked lots of thoughtful questions. He also started talking to Hannah's stomach, saying, "Hi, baby! Can't wait to meet you." What touched Hannah most was when he pulled out his guitar one evening and picked a soft soothing tune. "That's for Sophia," he said.

Hannah had been so used to living with anxiety, she wondered if feeling that way was simply a familiar habit. It was a useful emotion that had served her over the years. Perhaps it protected her from giving her heart to men who would not cherish it. Perhaps it nagged at her to go after her dreams. Whatever the case, she finally realized she didn't need it anymore.

• • •

On July 14, 2009, during a cool Seattle evening, Hannah made the final push that brought Sophia Grace into the world. When the delivery doctor showed her to Hannah, the newborn opened up her clear blue eyes and looked straight at her mother. "Oh my God! She's perfect!" Hannah exclaimed, then dissolved into sobs of joy.

Sophia belonged to her, and she belonged to this baby. As Hannah looked around the room and saw her beaming husband, stepdaughter, and stepson, she felt she belonged to them all.

Kelly

In the year and a half that Kelly had been trying to have a baby, she had endured three failed attempts to use her frozen eggs, one futile IVF round with her fresh eggs, one IVF pregnancy, and one miscarriage. Losing the pregnancy was by far the most devastating, but Kelly found a small bit of comfort in knowing her body could at least conceive a baby. As the days passed, the fresh grief was replaced by a growing faith that she could do it again.

Kelly and Dan had decided that if their last IVF round failed, they would consider using donor eggs. They had already met with their fertility clinic's counselor, who told them that many of her patients wished they had used donor eggs earlier instead of wasting years and money trying to have a biological child. She insisted that once they were pregnant, they often forgot their baby had another woman's DNA. In the meantime, she had started to see Dr. Andrew Toledo, an infertility specialist at Reproductive Biology Associates who specialized in treating older women and also advocated donor eggs, which had a nearly 70 percent success rate, compared to less than 5 percent with Kelly's forty-four-year-old eggs—not to mention the dramatically reduced risk of birth defects.

But now that Kelly had experienced being pregnant, she had a harder time giving up her dream of having a baby with her own genes. Even though she was near the absolute age limit for having a biological baby, her ob-gyn encouraged her to try her fresh eggs again. So did Dr. Toth, who prescribed more oral antibiotics to prepare her uterus for another transfer. When the analysis of her miscarried fetus showed no signs of chromosomal abnormalities, Dr. Toledo gave her the go-ahead to play the IVF game one more time.

A few months after her miscarriage, Kelly prepared for her last IVF round by scheduling the arduous series of doctor's visits to retake all her tests. In the meantime, she and Dan also followed Dr. Toledo's advice to apply to the clinic's donor egg program, since the matching process could take several months. Although RBA offered frozen eggs, he recommended using a donor's fresh eggs, which drove up the price of IVF from $15,000 to $25,000 but had a better shot of being fertilized with Dan's aspirated sperm sample, which contained fewer swimmers than what he would have produced prevasectomy.

Kelly was so focused on her tests that she didn't think much of the application until a few weeks later, when she received an email from RBA's donor egg coordinator with the words "FRESH MATCH" in the subject line. It read, "Dr. Toledo approved a donor for a fresh match. . . . You have 7 business days to make a decision regarding donor # FJ95890. FYI: This donor is proven. She donated for a fresh cycle in 2008 and the recipient is pregnant!!"

Kelly wasn't sure what to do. She wasn't ready to make a decision about using donor eggs, let alone select a donor. She hesitated before opening the attached profile of the potential genetic mother of her child. Part of her was sure she'd be disappointed, that the donor wouldn't have enough in common with her. But another part was more worried she'd like her.

Indeed Kelly was impressed with donor # FJ95890, a twenty-one-year-old married college student from Tennessee. She was blonde and fair-skinned like Kelly, and they both loved running, gardening, and traveling to London. She was a devout Christian, exercised regularly, played the piano, and enjoyed country music, cats, and chocolate. Kelly also liked the way the donor described herself as "easy going, outgoing, motivated, goal-oriented, detailed, happy, humorous, and witty." These were characteristics Kelly thought were important to be successful in life.

RBA had a policy of offering only anonymous donors, and Kelly briefly wondered if she and Dan should check out other agencies that provided clients with current pictures of the donors as adults or even the donors' children. But Dan felt strongly that he didn't want to have a current image of the donor in his mind. Kelly was content that RBA distributed donors' baby pictures. When she looked at a photo of her donor as a sweet little girl dressed in a white jumpsuit with a Peter Pan collar and white patent-leather Mary Janes, she felt a maternal pang. She knew she could love this child.

She and Dan weren't sure they would find anyone better, or how long it would take the clinic to suggest another match, and they didn't ask to see a second donor. Dr. Toledo explained that they could expect the donor to produce fifteen to twenty eggs. The clinic would fertilize them all, and she and Dan could freeze any leftover embryos in case they wanted siblings.

Kelly had to make a decision. Sometimes she felt that after all this time and money it was absurd *not* to try again for a biological baby. Despite the horrendous odds, she had proof her eggs were good enough to produce a pregnancy. What was one more try? She could withstand another round of drugs and another emotional rollercoaster. She and Dan could take out a loan for another $15,000. But she suddenly didn't want to waste any more time. Even if donor eggs worked, she'd still be forty-five by the time she gave birth to her first child. Another failed IVF cycle meant she'd be even older.

Kelly didn't want to wait any longer. She had been so inspired by being pregnant that she wanted to experience that feeling again as quickly as possible. The counselor had said it was normal for couples to go through a period of genetic mourning before making peace with donor eggs. She wondered what it would be like to carry and deliver a child that was made from Dan's DNA but not her own. She was surprised at how quickly and closely she had bonded to his twin boys. With donor eggs, she'd still be able to walk around with a big stomach sticking out and grow a baby in her body. By the following year, she had a 70 percent chance of having a baby in her arms. She told the donor coordinator, "We want to go ahead with this match."

Donor # FJ9580 produced nineteen eggs, which were injected with Dan's sperm and grew into eight embryos. Six embryos were frozen, and two were transferred into Kelly's uterus. Just as everyone expected, she became pregnant.

Nearly two months into her pregnancy, Kelly didn't have many classic symptoms. However, she didn't let herself worry. Donor eggs were supposed to be foolproof, and each ultrasound confirmed the baby was developing beautifully. But during lunch at a restaurant, she felt a sudden gush of wetness between her legs. Leaving a pool of blood on the chair, she ran to the bathroom and passed two clots. Dan consoled her on the phone that the blood loss didn't necessarily mean she was having a miscarriage. Shaking with nausea and terrified she would faint, she tied a sweater around her waist and managed to drive herself to the hospital, which was a half-mile away. The doctor did a vaginal ultrasound and reassured her the baby's heart was still beating. Kelly sank into the examining table with relief. Her seven-week-old baby was still alive.

But by the next day, the heartbeat had stopped.

In a fog of despair, Kelly tried to remind herself that young women with good eggs have miscarriages too. But she was becoming convinced something was wrong with her body. The worst part was that no one knew how to fix it.

They returned to Reproductive Biology Associates the following month. In her seventh transfer during the two years she had been trying to have a baby, two of the remaining six embryos were thawed and transferred to her uterus.

Two weeks later, Kelly heard the nurse announce over the phone, "You have a positive pregnancy result." She loved hearing those words, but she quickly reminded herself that her challenge was *staying* pregnant. She had to remain cautious until the nurse called back a couple days later with results of another blood test. "Oh my goodness!" the nurse exclaimed. "Your numbers have more than doubled." In the early stages of a pregnancy, the pregnancy hormone human chorionic gonadotropin doubles every two to three days. Overjoyed, Kelly embraced Dan and buried her face in his chest.

At the vaginal ultrasound a couple weeks later, Kelly scrunched her eyes shut and squeezed Dan's hand. "You look," she told him. By now, he had plenty of experience finding the black dot on the computer screen that indicated a gestational sac was growing.

"Here it is," the technician said. Kelly opened her eyes and looked up at Dan. Tears flowed freely down their cheeks, and they high-fived each other.

The next six weeks were the most anxious of her life. Every week she underwent a vaginal ultrasound to check on the baby's progress. "Could you please find the heartbeat first?" she directed the technician, asking her to take the other measurements after she delivered the news that the baby was still alive. As soon as Kelly heard the familiar whooshing sound of the fetal heartbeat, she relaxed.

She rejoiced when she felt nauseous or was so tired she wanted to go to bed by 8 p.m. Those symptoms reassured her that she was really pregnant.

Finally they made it to the end of the first trimester. "You're gonna make it this time," Kelly told herself. Sometimes she even forgot that the baby swimming around in her stomach didn't have her DNA, except when she wondered what she would write in the section of the baby book asking, "What are the mother's traits?" The counselor was right: it didn't matter. This was her baby.

They told the twins first. "You're going to be big brothers," they announced one evening after dinner. Then they showed them the ultrasound pictures.

Kelly was bursting to share the good news and offered to host her brother's fiftieth birthday party. When their guests arrived, they gathered outside to play football and roast marshmallows and oysters in metal fire pits on the side porch. An hour later, she eagerly rounded up everyone to pose for a photograph on the steps of the old Victorian. She set the self-timer button on the camera and ran back to join the group. "Okay, everyone, on the count of three, say 'Kelly's pregnant.'"

Everyone lost their composure as they tried to figure out if they had heard her right. The camera flashed.

Then they all surged toward Kelly, hugging and congratulating her.

They tried the photograph again. Kelly set the timer and took her place between her husband and bonus sons. The camera flashed.

Sarah

Paul and I had been soldiering through the summits on and off for well over a year. We had ditched the books long ago. Our latest module consisted of hanging out with other parents to learn more about what it would actually be like to have family. In addition to babysitting my niece, we spent lovely afternoons in Connecticut meeting the children of his friends. Paul set up happy hour drinks with his dad buddies.

Once again we came away with vastly different impressions. While I admired the rows of children's school pictures in the hallways, Paul saw how their school schedules would restrict our travel options. While I learned of stimulating afternoons filled with children's activities, he heard about boring afternoons filled with children's activities. I saw a richer life. He saw a lifetime responsibility.

Paul said he was touched when his friends told him how much they loved their children and enjoyed being fathers, but he was troubled by their accounts of how much work kids were. A couple of his friends confessed they looked forward to their kids going away to college. Other friends talked of debilitating divorces, challenging teenagers, and overwhelming pressure to support their family during a difficult job market. Yes, parenting was rewarding, they said, but it was also exhausting. They experienced higher highs—but also lower lows. "You've really got to want it," they told him.

Paul's position became crystal clear. He said he would give me anything in the world: We could get married. We could volunteer with children. We could even move to San Diego to be closer to my niece. But he absolutely didn't want children of his own.

Suddenly there were no more summits, no more *someday*. My "maybe" man had officially become a "no" man.

We saw a therapist who listened to Paul's list of reasons for not wanting children. I had hoped we could mine deeper issues: Was he scared that a baby would be a rival for my attention? That he'd lose me? Was there something from his childhood we hadn't explored? Many men had doubts but eventually became fathers. Was there some issue I was overlooking? Did he want to hear about my creative solutions to help him bond with the baby so he didn't feel left out?

Paul repeated his case: he valued his free time, didn't want the responsibility, and envisioned a rock-solid family of two and thought there was a real chance I would sign on. "There are many ways to be happy in life," he said. After a few more sessions of the same old song, the therapist said there was nothing to work on. The choice had become mine to make: Did I want to stay or go? I decided to see another therapist on my own and booked an appointment with Joann Paley Galst after reading her article in the *Ladies' Home Journal* titled "I Want a Baby, He Doesn't," in which she suggested that reluctant men often change their minds.

During our session, Dr. Galst told me her husband hadn't wanted children but overcame his hesitation once she agreed to go out a certain number of nights a week together to preserve their marriage. Although she hated leaving her son with a babysitter, she admitted that it was a good concession for maintaining a strong relationship, especially now that her son was grown.

She conceded that although it was possible Paul might change his mind later, he had been quite clear. I had to accept his answer. I had to decide whether to stay in my relationship with him and never have children; look for someone new who was willing to have children; or have them on my own. I felt overcome with the familiar feeling of dread. If I wasn't clear about my choice, she said, I could try an exercise: to secretly make a decision and stick with it for a couple of weeks to see how it felt.

I had plenty of practice visualizing being a mom, so I imagined not having children. Although I was now more in touch with my baby longings than ever before, I was also aware of my ambivalent streak that had let so many important years slip by. I knew it wasn't a complete accident that I had fallen in love with a "maybe" man.

So I imagined myself footloose and child-free. When I was tired and wanted to go to bed early, I was relieved I didn't have to wait until my kids were in bed first. When I was waiting in an airport, I was happy I didn't have to chase around a toddler. When Paul and I were debating who should clean up after dinner, I was grateful we didn't have to argue about who would bathe, dress, feed, and rock a baby. When we met friends for drinks after dinner, I was thankful we didn't have to rush home to pay the babysitter. I liked working late into the evening. I liked dressing up and meeting friends for drinks. I liked going to the gym whenever I wanted. I liked going to wine bars on Sunday afternoons. I liked sex. And sleep.

Paul and I *did* enjoy our family of two. We could organize some apartment swaps in foreign cities. We could take cooking and Pilates classes. Visit more museums. See more plays. Become neurotic dog owners. Start a nonprofit organization that helped children. Host dinner parties with our child-free friends.

One evening after dinner, Paul and I were knotted on the couch listening to jazz while it stormed outside. I felt such warmth and love in our little cocoon. I couldn't imagine ever needing anything more.

I tried to remember the reasons I wanted children. Had I simply been scared of becoming one of the lonely old unmarried women in my apartment building, who trapped people in the lobby with needy conversation? I wouldn't be; I had Paul. I couldn't depend on children to take care of me in old age anyway. They grew up and left home. It was more important to have a good partner.

I wondered how many women who had frozen their eggs would never use them. Maybe they had frozen so they didn't have to commit to the decision of not wanting children. If they waited long enough, did they simply lose the baby urge, sort of like waiting out a cookie craving? Once they reached a certain age, did the biological clock finally shut up?

But after two weeks of child-free fantasizing, the longings came back full force. I thought of everything I would miss: hanging Pilgrim art on the refrigerator and growing bean shoots in egg cartons in the windowsill; making up songs about boogers and singing them in the car. I remembered spending an afternoon with Kate in Target; we walked around for an hour smelling lip gloss and eating animal crackers.

I started playing a little game with myself to see how a baby would fit into our lives. Throughout the day I asked myself, "If we had a baby right now, what would we be doing?" Could I still go swimming? Could we take her to brunch? When I woke up to use the bathroom at 5 a.m., I thought, "If I had to get up right now and couldn't go back to bed, would I mind?" On Sunday nights when Paul was working at his computer and I had finished watching *Big Love*, I wished I could help a child get ready for bed: chasing her down the hall after a bath, reading her stories or negotiating which stuffed animals she could have in her crib. I imagined Paul and me sneaking in to check on her while she slept, swelling with pride that we had made such a beautiful child together. The child-free books promised a life of order, but I was starting to choke on all our peace and quiet. I craved noise.

I didn't want to miss out on everything for fear of sacrificing my free-dom or enduring a few sleepless nights. I wanted it all: sullen teenagers, tantrum-throwing toddlers, social studies homework, messy schedules, and babies crushing Cheerios on the floor. I'd been on enough trips. I'd sat through enough long dinners. I'd slept in enough weekends. I knew that even with kids, I'd still be able to do those things sometimes. I knew the years would fly by, and I would have decades left for more child-free living full of convenience, condos, and cheese plates.

Sometimes I felt hurt that Paul couldn't tolerate the inconvenience of one little baby, even if it meant losing me—that he would rather have none of me than share me with a child. But I realized it had nothing to do with me. Even though Paul would be an amazing father, he didn't want to be one. Just as I couldn't argue myself out of a need, I couldn't convince him of a desire. And I had run out of years to try.

Sometimes I wondered whether I was choosing a baby who did not exist over an actual man who was here now and made me deliriously happy. It felt as if I were choosing which limb to cut off. If I thought about losing Paul, it hurt. If I thought about losing Claire, it hurt. But I knew I didn't really have a choice. If I stayed with Paul and didn't honor my desire to have a child, I would grow to resent him. No experience with him could win in that kind of contest, and no requited love could survive unrequited baby hunger; in the end, I'd have neither husband nor child.

And just like that, I made a decision: I wanted a husband *and* a family. The baby panic rushed back, but this time it wasn't the old hyperventilating variety. It was the helpful, encouraging kind that said, "If you want this, girlfriend, then you better go get it."

And so one Sunday evening I told Paul what I had been avoiding for more than three magical years: I needed to go after what I wanted.

Because he loved me, he let me go.

We both sobbed and clung to each other for a long time. It was the worst kind of pain—for Paul, for love, for love I was scared I would never find again, and for the child we would never know.

I grieved our seven-year-old boy with Paul's matching glasses, curly hair, and infectious laugh, who loved his mother as much as his father did. In my fantasies, they hung out in the kitchen and pestered me for tastes of what I was cooking. I grieved our three-year-old girl who shared Paul's knack for facial expressions. Paul would call me on his way home from a meeting and say he couldn't wait to see "his girls." Sometimes I imagined us crossing Central Park in a taxi on a clear winter evening. We were going home from a doctor's office with our baby's ultrasound pictures in my bag. He would rest his hand on my stomach as we watched the Manhattan skyline and imagined all the possibilities.

I had to let go of it all.

I spent the next couple of months doing the things women do after breakups. I woke up at 5 a.m. seized with ache and went to sunrise yoga. I painted the hallway of my polenta studio a beautiful cranberry color and hung up pictures of my family and friends (including one of Paul). I bought a Crock-Pot and made gallons of comforting fall soups. I bought new relationship advice books and wrote lists of what I had to offer a future partner. I took salsa lessons.

In San Diego for the holidays, I went shopping with my mom for new boots and date outfits of delicate cardigans with pretty lace camisoles. I got professional photos of me sitting on the rocks at La Jolla Cove. I wrote a new online dating profile with the headline "A bit of California sunshine to warm the winter chill." On the question of whether I wanted kids, I marked "Definitely."

I bought a six-month subscription to Match and posted my profile on a Saturday morning. When I signed on later that evening, there were twenty emails waiting.

Epilogue

So, does egg freezing improve women's lives, after all? If we look at the end result—whether those frozen eggs turn into babies—the answer is yes for only one of the three women profiled in this book who thawed their eggs. This success rate is in line with doctors' estimates that women have a 30 to 40 percent chance that the procedure will work. Although these women froze their eggs when the technology first became available and wasn't as effective as it is today, they all still expected their eggs to work. Yet two of three women did not get what they paid for.

But the question is more complicated than that and can't be answered objectively. Their faith that egg freezing *would* work set in motion positive events in their lives. They enjoyed years with less baby panic, comforted that they were getting a second chance at motherhood.

Both Hannah and Kelly believe freezing their eggs gave them time to find their husbands. Although freezing didn't dampen Monica's urgency to have a baby with Adam, it enabled her to feel more relaxed as she pursued subsequent relationships. It also gave her peace of mind as she focused on her endometriosis treatment. As for me, egg freezing gave me time to explore and enjoy a lovely relationship and become ready to be a mom.

I've also been able to calmly enjoy dating as I search for that curious mix of attraction to and chemistry with someone who shares my goals. Since breaking up with Paul two years ago, I've met dozens of interesting men, including one from San Diego with ten-year-old twins who said he'd love to have more children with me. Some seemed uncomfortable when I told them that I had frozen my eggs. Others thought it was the best idea they'd ever heard. One man told me over dinner, "You seem really chill about it all. It's so nice."

I'll also admit I've been laughed at (by everyone from strangers I've met on planes to family members with teenagers), as if expecting to find love and babies at age forty-one is the craziest idea in the world. The funny thing is that when I was younger, I was terrified of being in this boat at this age. I could never have anticipated that the boat would get a lot more fun. Or that I would feel prettier and softer. I no longer wish that everything would work out; I've simply made up my mind that it will.

Then it did, just as it always does when you keep an open heart and go on enough Match dates. I met a wonderful forty-five-year-old single dad from New Jersey who wants more kids and wanted to hear all about my frozen eggs. Four hours after meeting at an Upper West Side wine bar, we were making out in Central Park in a warm September foggy mist. A month later, he said the sweetest words I have ever heard: "I can't believe I found you."

I can't stop kissing him.

Egg freezing allowed me to change the narrative of my life from mourning and desperation to hope and potential. I wake up every morning feeling good about my future. That is worth something, even if those eggs don't give me biological children.

Perhaps the better question is this: Are these psychological benefits worth taking the risk that you might not have a baby with your own DNA?

Some doctors have argued that egg freezing serves a valuable purpose by helping women get in touch with their desire to become mothers. When a woman freezes her eggs, two things happen: she comes to terms with the fact that her fertility is fading, and she invests significant time, energy, and money in protecting that asset. The combination is a powerful catalyst. "You're making a public declaration to have a baby," explains Dr. Jamie Grifo, a fertility doctor who helped start the egg-freezing program at New York University. "You have made a statement and taken action. You own the decision. You're not being a victim." In fact a survey of 240 women who visited a New York City fertility clinic for a consultation on egg freezing between 2005 and 2011 found that 84 percent said they had discussed with family their intention to undergo the procedure, and 78 percent reported talking about it with friends.

Georgia Witkin, the psychologist I had seen before my first round of egg freezing in New York, said there's something powerful about having those eggs in the freezer. I think there's something more powerful about *putting* those eggs in the freezer. Paul had awakened my longing to have children, but the act of freezing made me commit to it. In my case, eight times.

Critics' biggest concern is that women would use egg freezing as an excuse to ignore their biological clock. Yet surprisingly, the women in this book believe the act of freezing motivated them to take steps that brought them closer to becoming moms: Hannah, Kelly, and Monica became more serious about dating, and I left someone I dearly loved to be free to find a partner who also wanted to be a father.

It's also important to point out that these women didn't use their frozen eggs as an excuse to wait indefinitely. As soon as Kelly was married, she started trying to have a baby. Monica was so aware of her ticking clock that she decided to start the journey without a partner. And Hannah, worried about her creeping age, finally pushed her husband to make a decision to have more children.

Egg freezing doesn't silence the biological clock. Rather, it temporarily dulls the ticking so you can catch your breath and make good life choices. After a while, the noise, now quieter and steadier, faithfully reappears. But instead of feeling like a victim paralyzed by anxiety, you feel more in command of your own destiny. It is that mindfulness that makes me do what I'm supposed to do to make my life go in the direction I want. It forces me to put on a dress and a smile and attend another alumni cocktail on the nights I would love to stay home and watch *Mad Men* in my yoga pants. It makes me find the enthusiasm to tell yet another Match date that I love shopping at farmers' markets on Sunday afternoons.

As for ending up empty-handed, Kelly found a way to have a baby using donor eggs, and Monica likely will consider that option in the near future. Of course, donor eggs are not easy or affordable options for all women whose frozen eggs won't work. Yet as egg banks pop up across the country, the price is sure to come down.

But I am writing only about my experiences and those of a handful of women. Dr. Grifo, whose egg-freezing program has accepted more than one thousand women over the past eight years, says that while he's met women who undergo the procedure and then feel motivated to find a partner, he's also known others who freeze their eggs to "check the box" but have no intention of ever using them. Others freeze their eggs as a way to figure out relationships or as a form of divorce therapy and eventually get pregnant with another partner naturally.

Two lessons are clear: there is no typical Egg Freezer, and the act of freezing has a different value for everyone, regardless of whether or not it works.

When Christy Jones envisioned offering egg freezing to the masses in 2004, she thought thousands of women would jump at the opportunity. No organization or governmental agency tracks the number of women who freeze their eggs, but according to rough estimates, fewer than five thousand women have undergone the procedure for nonmedical reasons. It's hard to know why. More than 40 percent of college-educated women between thirty-three and forty-six are childless, according to a 2011 study by the New York–based think tank Center for Work-Life Policy. Were women waiting for the economy to improve? More babies to be born? Were they waiting until their friends did it? Or until their ob-gyns recommended it? Jones suspected that more often than not, the women succumbed to inertia—the same kind that made them not take their fertility seriously in the first place. Perhaps they were hopeful they would still meet someone in time.

Yet many doctors are reporting that their patient loads are starting to pick up, especially as egg freezing makes inroads into pop culture. In the spring of 2012, *Modern Family* television actress Sofia Vergara, thirty-nine, said she planned to freeze her eggs, and *Extra* cohost Maria

Menounos, thirty-three, went public with her decision to undergo the procedure the year before. Now that thirty-one-year-old Kim Kardashian has announced that she's joining the club, the transition of egg freezing from desperate to trendy is complete.

And more clinics are offering the procedure. A 2009 survey by the University of Southern California found that 51 percent of U.S. clinics now freeze women's eggs—65 percent of which offer the service to Clock Tickers—and 25 percent more plan to offer it in the near future. According to the report, 337 babies have been conceived in 857 thaw cycles in the United States.

This widespread adoption has had significant consequences for Jones's business model. A few years ago, Extend Fertility's partnerships proved as fragile as the eggs they were freezing. Some clinics had too many doctors and locations, and it was hard to streamline care. Some had too few clients. Some doctors decided they didn't need a third party to run their egg-freezing program. "Patients already know our name," explained one doctor. "Why would I need you?" They already froze and stored sperm and embryos in their labs; it didn't make sense to outsource for eggs.

In the summer of 2010, Jones (who has since married her longtime boyfriend and had a baby) changed Extend Fertility's business model. Rather than try to convince doctors to let her company handle storing eggs, training staff, supplying the freezing solution, and billing clients for a fixed cut of each patient's fees, she decided to focus on patient education. Extend would host events, distribute newsletters, and serve as a clearinghouse for information about egg freezing and refer clients to clinics it endorses. For example, in an effort to make the topic seem less formidable to potential patients, Extend recently hosted a spa event in Seattle in which women walked through theme rooms to learn about Botox, facial fillers, and fertility. Over time the website would only list centers that had experience making babies from frozen eggs, and the clinics would pay an annual fee to be included on the site. By the beginning of 2013, Extend's network had grown to ten partner clinics, including several whose doctors had once snubbed Jones.

In perhaps the biggest endorsement yet of egg freezing, in October 2012 the American Society for Reproductive Medicine declared that the newer vitrification technique should no longer be considered experimental. The latest opinion cited several studies that IVF pregnancy rates using these eggs are similar to those using fresh eggs and that babies born from frozen eggs show no increase in chromosomal abnormalities, birth defects, and developmental deficits than the rest of the population. Yet the opinion limits the endorsement to only younger women and still discourages offering egg freezing to Clock Tickers: "Data on the safety, efficacy, cost-effectiveness, and emotional risks of elective oocyte cryopreservation are insufficient to recommend elective oocyte cryopreservation. Marketing this technology for the purpose of deferring childbearing may give women false hope and encourage women to delay childbearing."

In the meantime, the science continues to improve, especially as more clinics embrace vitrification. Other efforts are in the works to make egg freezing more effective, including new testing methods to determine how many eggs a woman should save. Currently embryologists freeze the eggs that look good under a microscope; however, more than half of those eggs could be chromosomally abnormal, according to Dr. Geoffrey Sher, a Las Vegas–based fertility doctor who runs eight clinics across the country. He has started offering a testing service known as comparative genomic hybridization. For $3,500, scientists can extract a polar body from an egg and study its chromosomal structure. Critics say the technique is time-consuming, expensive, only partially accurate, and potentially damaging since it involves cutting the egg. But until a better method comes along, it might help give women a clearer picture of how many viable eggs they are banking and whether they should pay for additional cycles.

The problems that have plagued the industry for the past decade remain. The vast majority of practitioners who hang up egg-freezing shingles have yet to thaw significant quantities of eggs. In a 2009 survey of American clinics that offer egg freezing, nearly half had never thawed a client's frozen eggs, some thirty clinics reported no live births from the eggs they had thawed, and eleven achieved only one live birth. A small number of clinics are responsible for the majority of babies born from frozen eggs: only eight clinics reported achieving ten or more births.

There's still a lack of data in the field. The Centers for Disease Control recently started collecting egg-freezing statistics from clinics, but has yet to make them public. Some scientists have been censured for refusing to publish their results in peer-reviewed publications or share their freezing recipes. And no one knows the long-term health of children conceived from frozen eggs. The pharmaceutical company EMD Serono is partnering with Fertile Hope, a nonprofit organization that helps cancer patients learn about fertility preservation, to establish a registry to track four hundred egg freezers and document their outcomes and the health of their babies.

The hope is that the removal of the experimental label from egg freezing will encourage more doctors to study it and more ob-gyns to recommend it to patients when they are young and their eggs are more viable. The increased acceptance might spur doctors to compete for patients by publicizing the tallies of babies born from their labs. Maybe they would even reduce their prices. Women would become more discriminating consumers and hold doctors accountable. The ASRM could reward high-performing clinics by designating them centers of excellence. The whole field would improve.

But in the future women may not need to freeze their eggs, as researchers investigate even more extreme methods of fertility preservation. One new procedure involves surgically removing and freezing ovarian tissue and later reimplanting it in the hope that a woman's body will resume ovulation of those younger eggs. The technique is being developed for cancer patients who are too young or too sick to undergo the hormonal stimulation required for egg freezing, and it isn't recommended for older women, who have fewer eggs to work with. Although the procedure has produced only a couple dozen babies so far, it might be a useful option for Clock Tickers down the road.

Finally, scientists have been experimenting with ways to rehabilitate old eggs. Earlier in this decade, researchers reported achieving a pregnancy by removing the DNA center from a woman's fertilized egg, inserting it into another woman's egg from which the nucleus had been removed, and transferring the resulting embryo to the first woman's womb. The woman became pregnant with triplets but did not carry them to term. In another example, several babies were born after scientists took older women's eggs and injected them with the cytoplasmic fluid from younger women's eggs to nourish them. Both research projects have been banned or put on hold because they resulted in creating embryos with three biological parents, a potential legal quagmire. But they could result in new fertility therapies to compensate for low egg quality.

Despite all of these advances, the greatest potential reproductive revolution still awaits us. If scientists can figure out how to make new eggs from stem cells, a woman could have access to her own potentially unlimited supply of good eggs at any age. That day may not be too far off; a Chinese team recently discovered stem cells in the ovaries of adult mice that led to the creation of new eggs and eventually healthy offspring. Researchers from Harvard and Edinburgh University announced they had successfully made human eggs from stem cells and are now seeking the British government's permission to fertilize them. However, experts warn it will be several years before they can be used to create a baby.

Until then, practitioners such as Michael Tucker (who finally started accepting Clock Tickers in 2009) say that the real value of egg freezing isn't helping women save their own eggs. It's using the technology to transform the practice of donor eggs and make them more affordable to thousands of infertile women. Tucker has helped create a national network of donor egg banks, called Donor Egg Bank USA, that offers the frozen eggs of 130 young women. And several entrepreneurs have started similar businesses that allow patients to buy part of a donor's harvest, often at prices closer to traditional IVF cycles or for a fixed fee that guarantees a full refund if a patient doesn't deliver a baby after a certain number of tries.

For women seeking to freeze their *own* eggs, cost is still the biggest barrier. The procedure is mostly embraced by women who have the money or can shoulder significant debt. Or, as the *New York Times* recently reported, they have parents who are willing to foot the bill in the hope it will produce grandchildren. But perhaps generic fertility drugs will become widely available, or increased competition will force clinics to lower prices or offer discount packages, and the procedure will become more accessible. Internet ads currently beckon women to travel to India for egg-freezing deals. Or, as Christy Jones and others have speculated, egg banks may allow eligible women to participate in a unique exchange program: they freeze their eggs for free or at a reduced cost as long as they donate some to infertile women.

Perhaps, as in the case of two workers at the consulting firm Mc-Kinsey, women may ask their employers to pay for it. The company with the notoriously killer workload turned them down, but the women's request is a refreshing exception to feminist fears that employers would use egg freezing as a way to coerce employees into not taking time off to have children. These women, who were in their late twenties, weren't using egg freezing as a last-ditch effort to have a family but as an instrument of power that factored into their overall life plans. There are other indications that women are thinking of freezing their eggs earlier, even if by just a few years. An analysis of the women who received a counseling session about egg freezing at Reproductive Medicine Associates in New York found that the age of the average patient had dropped to thirty-seven in 2011 from thirty-nine in 2005.

This is a positive trend in terms of individual patients enjoying a higher chance of success. But we still must ask: Is egg freezing good for women, children, and society as a whole? Should it be regulated? And how comfortable are we as a society with pushing this reproductive boundary? Maria del Carmen Bousada, a single Spanish woman who gave birth to twin boys at age sixty-six in 2006, thought she would follow in the steps of her mother, who lived to be 101, but she died of stomach cancer two years later. And few people can forget the sad specter of Elizabeth Edwards succumbing to breast cancer in 2010 at age sixty-one, leaving behind an eleven-year-old daughter and a nine-year-old son.

Currently no U.S. law exists to restrict the age at which women can seek medical help to conceive children. Some states mandating insurance coverage for fertility treatments will not accept women over forty-five, but it's largely up to individual doctors to decide whom they will treat. Fertility clinics often set their own caps of fifty or fifty-five for female patients or require that the combined age of a couple be no higher than one hundred. (For example, if she's fifty-five, he should be no older than forty-five.) The reasoning: at least one parent will be alive long enough to see the kids to adulthood. Israel, one of the few countries to legislate such matters, allows women to use their frozen eggs until they are fifty-four.

In September 2011 *New York* magazine spared little subtlety on its cover, blaring "Is She Just Too Old for This?" and featuring a wizened, white-haired woman cradling her naked pregnant belly à la Demi Moore's famous *Vanity Fair* pose in 1991. "When a 50-year-old decides to strap on the Baby Björn," wrote the article's author, Lisa Miller, "the choice prompts something like a moral gag reflex." A significant percentage of the more than three hundred online comments agreed. Here's a sampling:

I'm sorry, but this is sick. Being brought OUT OF MENO-PAUSE to carry a child? No, just no.

It's called accepting the choices you've made in life and playing the hand you're dealt, instead of dreaming up some new Frankenstein-y method to get what you want, no matter the cost.

Too bad for the children, and the rest of us. We're all to be damned by these irresponsible adults that don't really think through the ethics, needs of the child, and the impact on others of such very, very poor decision making.

Nothing wrong with it at all, just be prepared for the possibility of your baby being retarded, dying before seeing them graduate college, and for every third person you meet to ask how much you enjoy being a grandparent.

53 and pregnant? People shouldn't even be having kids when they're 43. Selfish, selfish people.

Given the cover's shock tactics, you would think the numbers of women having babies after age fifty is huge. It's not. According to the Centers for Disease Control, 569 babies were born in the United States to women over fifty in 2009, the most recent year for which detailed data are available. The real growth is among women in their forties: in 2009 women forty-five to forty-nine had 7,320 babies, a number that has doubled since 1997. And women forty to forty-four accounted for 105,827 births, an increase of nearly 300 percent since 1997.

The biggest concern isn't for the child of your random fifty-three-year-old woman using her AARP discount card at Babies R Us. It's for the surge of first-time moms at forty-four who have a second baby at forty-eight. What are the consequences of *that* trend?

A recent study of 177 Israeli women over forty-five found they were three times more likely than younger women to suffer from high blood pressure and gestational diabetes during pregnancy. Another study suggested that half of women over forty-five will develop one of the conditions, especially as the progesterone given during IVF may cause blood pressure and cholesterol to temporarily spike. Although doctors routinely reassure women that these conditions can be successfully managed with medication, monitoring, and often forced bed rest, pregnancy is still no picnic for this age group. The increased stress on an older woman's heart may lead to cardiac complications, and the stress on the back will further hurt many older women's already weakened muscles. They are also at risk of their placenta blocking the opening to the birth canal, a condition known as placenta previa, which can cause vaginal bleeding and premature labor. Older women produce large amounts of estrogen, which one study found helped protect their skin, hair, bones, and blood vessels and ward off osteoporosis but can also accelerate the growth of breast cancer. Considering that most women are advised to avoid mammograms from the moment they're pregnant until they finish breastfeeding, the cancer can go undetected for several years.

Their babies also face a higher risk of being born premature and consequently suffering from developmental delays as well as neurological, digestive, and lung problems. (The Israeli study found that more than 20 percent of the post–forty-five group gave birth at less than thirty-seven weeks, compared to 10 percent of younger women.)

But those warnings, while sobering, aren't enough to slow the trend. Pregnancies are managed. Premies survive. It's the idea of older mothers *dying* on their children, even their young adult children, that's at the core of public disapproval. While older dads likely have younger wives who can pick up the pieces if they die earlier than they had anticipated, older mothers usually partner with men closer to their age or older or have babies on their own. But public perception might be shifting on this issue too, especially as an increasing number of reports link older fathers to children's schizophrenia, autism, bipolar disorder, dwarfism, autoimmune disease, and miscarriage, prompting the *Wall Street Journal* to run the headline "How the Guyological Clock Could Change the Dating Game." The author, Amy Sohn, muses whether men should freeze their sperm by a certain age, feel pressured to start their families earlier, or, at the least, have more empathy for their female peers.

There's some research suggesting that women who give birth after forty are more likely to live longer than women who have children at a younger age; however, it's not clear how the conclusion applies to women taking advantage of fertility science. Were their abnormally healthy reproductive systems important indicators of longevity? Older mothers are also more motivated to stay healthy so they will be around for their children. According to other theories, kids keep mothers more active and socially connected, factors known to extend one's life. Or are women who even consider motherhood at a later age a healthier bunch to begin with? Surely women who lack energy and vibrancy and already feel old at forty-five aren't the ones thinking about signing up for the job.

Even if older mothers live long enough to scoot their kids off to college, the issue brings up another uncomfortable question: At what age could *adult* children stand to lose their mother? Is it culturally acceptable for a woman who has a baby at fifty and dies at eighty to leave behind a thirty-year-old? According to the 2008 actuarial life table that's used by the Social Security Administration, that's not an unreasonable scenario. A forty-five-year-old woman today can expect to live thirty-seven more years; a fifty-year-old can expect to live nearly thirty-three more years. Future generations of older mothers can count on living a bit longer, since expectancies probably will inch up by the time they have children. Considering that those numbers are averages, surely women who have better health care, religiously munch on kale chips, swallow fish oil supplements, and practice yoga well into their sunset years will be around even longer.

Yet one consequence cannot be avoided: the increasing incidence of older motherhood will fundamentally change the experience of young adulthood as we know it. In the new reality, adult children might be on the hook earlier to care for sick or frail parents. Rather than bounce around careers, cities, and relationships like their peers, they will be forced to grow up more quickly.

The psychological impact is harder to measure and may even be less relevant with successive generations. Adults who were born to older parents in previous decades often complain of having felt painfully out of place because their parents didn't fit in with their peers' parents: their parents dressed differently, preferred fuddy-duddy music, didn't know pop culture references, and often were excluded from social events. Being the odd child on the block is less of a concern these days, though, when there's a critical mass of older mothers, many of whom are indistinguishable from their younger peers sporting the same yoga pants and hair highlights.

Biology naturally made sure that women couldn't have children past an age when they couldn't care for them. Since science has disabled this natural check, does it make sense for government or some other regulatory body to establish age guidelines to protect children's interests? Or should we leave it up to fertility doctors to be the gatekeepers?

Age seems like an arbitrary cutoff, especially since one's medical history (high blood pressure, family history of breast cancer, obesity) and lifestyle factors (smoking, exercise, stress management) have a dramatic impact on longevity. Perhaps it makes more sense to evaluate a woman's overall profile regarding her capacity to care for a child. Few people worry about a baby born to a forty-seven-year-old married woman who already has four children. It's assumed the "last of the litter" will be taken care of in such a large family. But doctors deciding whether to treat first-

time moms after fifty might consider the age of their partner in addition
to whether the child would have older siblings who could (and *would*)
step up. If the mother is using donor sperm, does she have brothers or
sisters or close friends who would assume responsibility in the event she
couldn't care for the child? Do these women plan to have more than one
child to give their kids a larger family?

It would be refreshing if there was more recognition of the goodness
of women having babies into middle age. As an antidote to the "children
as misery" literature, a 2011 study looking at the happiness levels of more
than 200,000 people from more than eighty-six countries found that
people over forty are happier with children than those without. In this
country, older dads are usually the ones portrayed as being able to enjoy
and appreciate parenting in a way they couldn't when they were younger.
They likely have more time and financial security and are more nurturing,
due to an age-related drop in testosterone. Coverage of older motherhood, on the other hand, plays up the lack of energy, the abrupt change
in lifestyle, the stigma of being mistaken for the child's grandmother, and
the alienation from younger mothers.

Yet just like older fathers, many older mothers believe their age has enabled them to be better parents, and studies show that the higher the age of
the mother of IVF babies, the better those kids perform on reading, math,
and language tests. These mothers are also more financially secure, patient,
and confident and more likely to be in a good marriage than younger mothers.

Older motherhood will undoubtedly continue to thrive and should
be celebrated and embraced. Yet as Dr. Eleonora Porcu argues, encouraging generations of women to rely on hormone injections and surgery
is not real progress for women. We need to have our children when
our body is most able, and it would help to live in a society with fami-
ly-friendly employers, affordable child care, and fathers willing to share
the work. We shouldn't have to turn to extreme science to rescue us.

Talking about egg freezing rather than preparing for infertility is
helping women already. When I told my friend Janelle that I had decided to freeze my eggs, she briefly wondered whether she should freeze
hers too, since she was thirty-four and had a family history of premature

ovarian failure. She was in between jobs, and her boyfriend of two years was juggling finishing his MBA while working full time. They had just moved in together, and she wanted a few years to settle into a routine and get married before starting a family. However, she was aware that she didn't have time to waste if she wanted several children. "This isn't the best time," she told her boyfriend, "but it's not the worst time either." He agreed, and she became pregnant two months later. For Janelle, learning about egg freezing, considering it, and ultimately deciding against it forced her to realize she would lose her fertility if she didn't take action.

It remains to be seen how egg freezing will be embraced by the medical community and future generations of Clock Tickers, but an important social experiment has already begun. Its very existence represents an important expansion of reproductive choice in how women perceive and manage their fertility.

For women who don't have other options, egg freezing is the best tool we have right now, and women should be given access to good doctors and good science. An opinion piece in the February 2012 issue of *Fertility and Sterility* suggests that age-related infertility should be regarded as a medical problem and that egg freezing should be considered "preventative medicine." The author insists that doctors have a responsibility to discuss fertility preservation with their patients and make sure that young women truly understand all their options. That is the best hope for changing the most recent U.S. Census statistic estimating that one in five American women age forty to forty-four is childless. Half of those women wish they could have children. Given those figures, nearly 10 percent of the post-forty female demographic could have benefited from having access to egg freezing just a few short years ago.

We are witnessing an unprecedented time in history when women's rapid social and career advancements have collided head-on with their biological clocks. Women have enjoyed an expansion of choice in nearly every area of their lives, except in their ability to have children. We undoubtedly will be trying to navigate this mismatch for generations to come, but if technology can temporarily compensate by adding another layer of choice, that is a reprieve indeed.

Acknowledgments

My most sincere thanks to John Jain, MD, medical director of Santa Monica Fertility and Egg Freezing Center, for reviewing the manuscript in its entirety and offering thoughtful comments.

A writer who has a good editor is extremely lucky. I, however, hit the jackpot twice. Thank you to my first editor, Kerri Kolen, who understood that this book was about more than egg freezing. I am grateful for your big heart and sensitive editing and for showing me how to mine the power of narrative. Thank you to my second editor, Priscilla Painton, for your remarkable eye for things big and small and for always finding a way to make this book just a little bit better. I appreciate your enthusiasm and generosity. Thank you to David Rosenthal, for taking a risk on me in the first place and for coming up with the title. Thank you to Jonathan Karp, for setting a new industry standard in creative thinking about ways to connect readers to books.

I want to thank other folks at Simon & Schuster who have shown me what a delight it's been to work with a first-rate publisher: Michael Szczerban, Sydney Tanigawa, Elisa Rivlin, Phil Metcalf, Judith Hoover, Emer Flounders, Tracey Guest, Anne Tate, Marie Florio, and Nina Pajak. I am grateful for your careful attention, warmth, and encouragement.

Thank you to my agent, Paula Balzer, for your friendship, vision, and confidence. Thank you to Marty Munson, my amazing former editor at *Marie Claire*, for assigning me a pair of essays on egg freezing way back when. To my former Columbia professor Sam Freedman, whose seminar on book writing I had the privilege to take in grad school, thanks for showing me that it was possible to write a book and for reminding me over the years about the importance of and pleasure in telling people's stories.

To my alumni writers' group—Harry Bruinius, Jonathan Englert, S. Mitra Kalita, Alice Sparberg Alexiou, and Michael Bobelian—I am grateful for your camaraderie, perspective, and useful advice during all parts of the book-writing process. Thank you to Minal Hajratwala, Sparrow King, and John Brown for graciously reading early drafts; your feedback was invaluable.

Thank you to my friends, who listened to my stories, sent me news links, put me up on reporting trips, or simply reminded me how fortunate I was to have the opportunity to write a book about a topic I care about: Janelle Anderson, Lynne Arciero, Adeena Babbitt, Bill Berkeley, Steve Bland, Narisa Cougar, Andrew DeLorenzo, Nina DeLorenzo, Anna-Marie Filippi, Patrick Fisher, Jim Fonda, Mick Green, Kurt Hummler, Mark and Ilyssa Israel, Sandra and Bruce Heller, Christina Hoppe, Scott Jones, Evan and Ruth Katz, Steve and Kris Kester, David Lawrence, Miranda Leitsinger, Andrew Marker, Victor Morange, Alan Mowatt, Michelle Nasir, Nancy Neilan, Jan Nguyen, Nora Nicolini, Mehran Sahami, David Schuster, Nilanjan Sen, Josephine Stoff, Hetal Shah, Nancy Smith, Michelle Steiger, Karin Tamerius, Xuan Thai, Jessica Thaler, Vicky Vuong, Jamie Weyand, Philip Yanni, and Steve Zupp.

Thanks to my mom, Jennifer Stoff, for always being my biggest fan and for setting the bar for motherhood high. I can't wait to follow your example. Thanks to my father, Hugh Richards, and my stepfather, Michael Stoff, for the countless ways you make me feel special. Thank you to my great-aunt, Helen Clarke; aunt, Jill Waters; brothers, Jason Richards, Matt Richards, and Bryan Richards; sister, Lauren Stoff; sister-in-law, Stephanie Richards; and cousin, Rachel Herod, who make me feel proud to be part of a family. And thank you to my beloved niece, Katherine Richards, and nephew, Ethan Richards, who constantly remind me about the joys of having little ones in your life.

Thank you to Bruno Tedeschi. I can't believe I found you, either. You were worth the wait.

Thank you to Megan Griswold, a very special woman who shared her story, but ultimately was not included in the book. Your fortitude and zeal inspire me daily.

To the women in this book, thank you for sharing the most intimate details of your lives in the hopes that we may make these anxious years a little easier for those who follow us.

Finally, to Clock Tickers young and old, may we all find our own paths to motherhood.

Sources

BOOKS

Nicki DeFago, *Childfree and Loving It* (London: Vision, 2005).

Diana Dell and Suzan Eram, *Do I Want to Be a Mom? A Woman's Guide to the Decision of a Lifetime* (New York: McGraw Hill, 2003).

Elizabeth Gregory, *Ready: Why Women Are Embracing the New Later Motherhood* (New York: Basic, 2007).

Sylvia Ann Hewlett, *Creating a Life: Professional Women and the Quest for Children* (New York: Miramax, 2002).

Patricia W. Lunneborg, *The Chosen Lives of Childfree Men* (Santa Barbara, Calif.: Praeger, 1999).

Donna Wade, *I Want a Baby, He Doesn't: How Both Partners Can Make the Right Decision at the Right Time* (Brooklyn: Adams, 2005).

NEWSPAPERS

Theresa Agovino, "Former Tech Exec Launches Company to Freeze Women's Eggs," Associated Press, Sept. 16, 2004.

Dan Even, "Dead Woman's Ova Harvested after Court Okays Family Request," *Haaretz*, Aug. 8. 2011.

Elissa Gootman, "So Eager for Grandchildren, They're Paying the Egg Freezing Clinic," *New York Times*, May 13, 2012.

Judith Graham, "Some Hope to Put Their Fertility in Deep Freeze," *Chicago Tribune*, June 20, 2004.

Gina Kolata, "Successful Births Reported with Frozen Human Eggs," *New York Times*, Oct. 17, 1997.

Melissa Ludwig, "Turning Back the Biological Clock," *Austin American-Statesman*, July 25, 2004.

Fiona Macrae and Nigel Blundell, "Why Older Mothers Are Likely to Live Longer," *Daily Mail*, Feb. 25, 2006.

Shari Roan, "Fertility in Reserve," *Los Angeles Times*, Feb. 25, 2002.

Amy Sohn, "How the Guyological Clock Could Change the Dating Game," *Wall Street Journal*, Sept. 18, 2012.

Sylvia Pagan Westphal, "Baby Steps: At Fertility Clinics, a New Emphasis on Gentle Methods," *Wall Street Journal*, Mar. 23, 2006.

MAGAZINES

Joann Paley Galst, "I Want a Baby; He Doesn't," *Ladies' Home Journal*.

"The Happiness-Health Connection," HEALTHbeat, Harvard Health Publications, Mar. 24, 2009.

"Having a Baby after 40," *People*, Apr. 29, 2002.

Richard Jerome, "In the Bank," *People*, July 1, 2002.

Diana Kapp, "Ice, Ice Baby," *Elle*, Apr. 2004.

Jeffrey Kluger, "Eggs on the Rocks," *Time*, Oct. 27, 1997.

Lisa Miller, "Parents of a Certain Age," *New York*, Sept. 25, 2011.

Emily Nussbaum, "Mothers Anonymous," *New York*, July 16, 2006.

Don Sider, "Cooling Off Period," *People*, Nov. 10, 1997.

ONLINE REPORTS

American Society for Reproductive Medicine, "Age and Fertility: A Guide for Patients," 2003.

Center for Nutrition Policy and Promotion, U.S. Department of Agriculture, "Cost of Raising a Child Calculator."

Sylvia Ann Hewlett and Lauren Leader-Chivee, "The X Factor: Tapping into the Strengths of the 33- to 46-Year-Old Generation," Center for Work-Life Policy, Sept. 2011.

Gretchen Livingston and D'Vera Cohn, "The New Demography of American

Motherhood," Pew Research Center, Pew Social & Demographic Trends, May 6, 2010.

Gretchen Livingston and D'Vera Cohn, "More Women without Children," Pew Research Center, Pew Social & Demographic Trends, June 25, 2010.

Frank Newport, "Desire to Have Children Alive and Well in America," Gallup Poll, Aug. 19, 2003.

Pew Research Center, "As Marriage and Parenthood Drift Apart, Public Is Concerned about Social Impact," July 1, 2007.

Practice Committee of the American Society for Reproductive Medicine, "Essential Elements of Informed Consent for Elective Oocyte Cryopreservation," Oct. 2007.

Practice Committee of the American Society for Reproductive Medicine, "Mature Oocyte Cryopreservation: A Guideline," Oct. 2012.

U.S. Census Bureau, "Fertility of American Women Current Population Survey," June 2010.

Barbara Dafoe Whitehead, "Life without Children: The Social Retreat from Children and How It's Changing America," Rutgers Marriage Project, 2008.

ACADEMIC JOURNALS

Jeffrey Boldt et al., "Human Oocyte Cryopreservation: 5-Year Experience with a Sodium-Depleted Slow Freezing Method," *Reproductive BioMedicine Online*, July 2006.

Christopher Chen, "Pregnancy after Human Oocyte Cryopreservation," *Lancet*, Apr. 1986.

Brian Doss et al., "The Effect of the Transition to Parenthood on Relationship Quality: An 8-Year Prospective Study," *Journal of Personal and Social Psychology*, Mar. 2009.

Raffaella Fabbri et al., "Oocyte Cryopreservation," *Human Reproduction*, Dec. 1998, Suppl.

Debra Gook et al., "Cryopreservation of Mouse and Human Oocytes Using 1,2-Propanediol and the Configuration of the Meiotic Spindle," *Human Reproduction*, July 1993.

Debra Gook et al., "Fertilization of Human Oocytes Following Cryopreservation: Normal Karyotypes and Absence of Stray Chromosomes," *Human Reproduction*, Apr. 1994.

Debra Gook et al., "Intracytoplasmic Sperm Injection and Embryo Development of Human Oocytes Cryopreserved Using 1,2-Propanediol," *Human Reproduction*, Oct. 1995.

James Grifo and Nicole Noyes, "Delivery Rate Using Cryopreserved Oocytes Is Comparable to Conventional in Vitro Fertilization Using Fresh Oocytes: Potential Fertility Preservation for Female Cancer Patients," *Fertility and Sterility*, Feb. 2010.

Julianne Holt-Lunstad et al., "Married with Children: The Influence of Parental Status and Gender on Ambulatory Blood Pressure," *Annals of Behavioral Medicine*, Dec. 2009.

Allan Jensen et al., "Use of Fertility Drugs and Risk of Ovarian Cancer: Danish Population Based Cohort Study," *British Medical Journal*, Feb. 2009.

Bengt Kallen et al., "Malignancies among Women Who Gave Birth after In Vitro Fertilization," *Human Reproduction*, Jan. 2011.

Rachel Margolis and Mikko Myrskylä, "A Global Perspective on Happiness and Fertility," *Population and Development Review*, 2011.

John Mirowsky, "Age at First Birth, Health, and Mortality," *Journal of Health and Social Behavior*, Mar. 2005.

Nicole Noyes et al., "Oocyte Cryopreservation: Is It Time to Remove Its Experimental Label?," *Journal of Assisted Reproduction Genetics*, Feb. 2010.

Nicole Noyes et al., "Over 900 Oocyte Cryopreservation Babies Born with No Apparent Increase in Congenital Anomalies," *Reproductive Biomed Online*, June 2009.

Kutluk Oktay et al., "Efficiency of Oocyte Cryopreservation: A Meta-analysis," *Fertility and Sterility*, July 2006.

G. Palermo et al., "Pregnancies after Intracytoplasmic Injection of Single Spermatozoon into an Oocyte," *Lancet*, July 1992.

S. E. Park, et al., "Chromosome and Spindle Configurations of Human Oocytes Matured In Vitro after Cryopreservation at the Germinal Vesicle Stage," *Fertility and Sterility*, Nov. 1997.

Thomas Perls et al., "Middle-aged Mothers Live Longer," *Nature*, Sept. 1997.

Eleonora Porcu et al., "Birth of a Healthy Female after Intracytoplasmic Sperm

Injection of Cryopreserved Human Oocytes," *Fertility and Sterility*, Oct. 1997.

Briana Rudick et al., "The Status of Oocyte Cryopreservation in the United States," *Fertility and Sterility*, Dec. 2010.

Lisa Schuman et al., "Women Pursuing Non-Medical Oocyte Cryopreservation Share Information about Their Treatment with Family and Friends," *Fertility and Sterility*, Mar. 2012.

Geoffrey Sher et al., "Selective Vitrification of Euploid Oocytes Markedly Improves Survival, Fertilization and Pregnancy-Generating Potential," *Reproductive BioMedicine Online*, Oct. 2008.

Scott Stanley et al., "Communication, Conflict, and Commitment: Insights on the Foundations of Relationship Success from a National Survey," *Family Process*, Winter 2002.

Michael Tucker et al., "Birth after Cryopreservation of Immature Oocytes with Subsequent In Vitro Maturation," *Fertility and Sterility*, Sept. 1998.

Michael Tucker et al, "Clinical Application of Human Egg Cryopreservation," *Human Reproduction*, Nov. 1998.

Michael Tucker et al., "Preliminary Experience with Human Oocyte Cryopreservation Using 1,2-Propanediol and Sucrose," *Human Reproduction*, July 1996.

J. F. van Uem et al., "Birth after Cryopreservation of Unfertilized Oocytes," *Lancet*, Mar. 1987.

Yvonne White et al., "Oocyte Formation by Mitotically Active Germ Cells Purified from Ovaries of Reproductive-Age Women," *Nature Medicine*, Feb. 2012.

Dori Woods et al., "Purification of Oogonial Stem Cells from Adult Mouse and Human Ovaries: An Assessment of the Literature and a View toward the Future," *Reproductive Science*, Sept. 2012.

Dori Woods and Jonathan Tilly, "The Next (Re)generation of Ovarian Biology and Fertility in Women: Is Current Science Tomorrow's Practice?," *Fertility and Sterility*, July 2012.

Nicole Wyndham et al., "A Persistent Misperception: Assisted Reproductive Technology Can Reverse the 'Aged Biological Clock,'" *Fertility and Sterility*, May 2012.

Index

Menopause, 4, 29, 181, 250
  early, 9, 51, 52, 73, 140, 181
  having babies after, 250–54
  obsolete, 53, 250
  side effects, 180, 181–82
Menounos, Maria, 244–45
Menstruation, 29, 82–83, 89, 158,
    180, 209
  fluctuating cycles, 82
Mental illness, 26, 200, 252
Miami, 124
Micromanipulation, 48
Miller, Lisa, 250
Ming-Na, 102
Minimal stimulation, 209–10
Minneapolis, 18
Miscarriage, 9, 51, 52, 147, 192,
    215–16, 230, 233, 252
  rates, 215, 216
Mixed-race babies, 54
Modern Family (TV show), 244
"Mommy shock," 38
Monica, 241, 243, 244
  longings, 18–23
  men, 95–99, 112–18, 133–36
  reckoning, 217–26
  reprieve, 69–72
  time, 153–57, 179–83, 195–201
Montreal, 174–77, 208, 210–11
Moore, Demi, 250
Moore, Julianne, 25
Morning sickness, 192, 215
Motherhood, 6
  after fifty, 249–54

after forty, 73, 155, 191, 197,
    203–5, 219–20, 232, 247–54
ambivalence about, 81–82,
    162–64, 171–73, 177, 182,
    236–37
single, 7, 17, 32, 110–11, 164,
    196–201, 217–19, 222, 243
stepchildren and, 5, 103, 119, 120,
    129–31, 162, 165, 166–67,
    182, 194, 196, 227
Movies, 31, 221
Mukherjee, Dr. Tanmoy, 83, 90, 91,
    208–9
Multiple pregnancies, 52, 54, 66, 109,
    140, 159, 185, 209
  selective reduction, 159–60
Music, 27, 41, 154, 166, 167, 237,
    253

Names, baby, 6, 111, 227
Nausea, 192, 215
New England Cryogenic Center,
    174
New Jersey, 18, 19, 70, 97, 127, 217,
    242
New Year's Eve, 112, 131, 196
New York City, 1, 13, 66, 83, 97, 110,
    126, 149, 154, 199, 206, 209,
    211, 242, 243
New York magazine, 38, 142, 250
New York Times, 16, 53, 248
New York University, 147, 242
North Carolina, 24
Notting Hill (movie), 31

About the Author

Sarah Elizabeth Richards is a journalist specializing in health and medicine, psychology, and social issues and has written for more than two dozen newspapers, magazines, and websites, including the *New York Times*, the *Washington Post*, the *Financial Times*, *Elle*, *Marie Claire*, *Slate*, and *Salon*. She is a graduate of the University of California at Berkeley and holds master's degrees from the Graduate School of Journalism and the School of International and Public Affairs at Columbia University. She is a three-time winner of the Newswomen's Club of New York's Front Page award. She lives in Manhattan.